AF843771

The Origin *of* Individuals

The Origin *of* Individuals

Jean-Jacques Kupiec

Ecole Normale Supérieure &
INSERM, France

(Translated from the French by
Margaret and John Hutchings)

World Scientific

NEW JERSEY • LONDON • SINGAPORE • BEIJING • SHANGHAI • HONG KONG • TAIPEI • CHENNAI

Published by

World Scientific Publishing Co. Pte. Ltd.
5 Toh Tuck Link, Singapore 596224
USA office: 27 Warren Street, Suite 401-402, Hackensack, NJ 07601
UK office: 57 Shelton Street, Covent Garden, London WC2H 9HE

Library of Congress Cataloging-in-Publication Data
Kupiec, Jean-Jacques.
[L'origine des individus. English]
The origin of individuals / Jean-Jacques Kupiec.
p. cm.
ISBN-13: 978-981-270-499-3 (hardcover : alk. paper)
ISBN-10: 981-270-499-X (hardcover : alk. paper)
1. Developmental genetics. 2. Biodiversity. 3. Molecular evolution.
4. Individual differences. I. Title.
QH453.K8713 2009
571.8--dc22

2008049845

British Library Cataloguing-in-Publication Data
A catalogue record for this book is available from the British Library.

Copyright © 2009 by World Scientific Publishing Co. Pte. Ltd.

All rights reserved. This book, or parts thereof, may not be reproduced in any form or by any means, electronic or mechanical, including photocopying, recording or any information storage and retrieval system now known or to be invented, without written permission from the Publisher.

For photocopying of material in this volume, please pay a copying fee through the Copyright Clearance Center, Inc., 222 Rosewood Drive, Danvers, MA 01923, USA. In this case permission to photocopy is not required from the publisher.

Typeset by Stallion Press
Email: enquiries@stallionpress.com

Printed in Singapore by Mainland Press Pte Ltd

For
Penny, David, Lana

Foreword

When I was a student, I got on very well with my professor of genetics whom I greatly admired — an admiration which subsequently turned into affection. I had developed the habit of visiting him regularly in his office in the Institute of Molecular Biology on the Jussieu university campus (nowadays renamed as the Jacques Monod Institute). I believe he shared the pleasure I experienced from our discussions which often went beyond topics of biology.

This all took place towards the end of the '70s, in an ambience that is difficult to describe nowadays. I had started my career very badly (it has not really improved much since). After refusing a position as a lecturer and giving up my PhD thesis, I severed all links with the academic world. I lived by casual jobs, day to day. I had only a few friends, none of whom were biologists or scientists: they were mostly aspiring artists and we led a life without any clear direction, occupied by never-ending discussions and walks around the streets of Paris.

Jean Tavlitzki was the sole researcher with whom I had kept in contact. A few years earlier I had attended a course he gave which enthralled me. One of his lectures concerned cell differentiation and that was the origin of my infatuation with the subject.

I am mentioning this because the starting point of this book comes from one of our many discussions. At that time I still thought that the development of an embryo was directed by a genetic programme which functions due to specific proteins. One day, he pointed out to me that no such proteins had ever been discovered in organisms with several cells. I was more than somewhat disturbed by this as their existence seemed so obvious to me that it could not be doubted.

Jean Tavlitzki had not said this to contest the predominant ideas of the time or to throw me off balance. It was simply an objective observation made with scientific precision. He did not imagine that in the future these regulators would not be discovered and went no further in drawing any kind of conclusion that could challenge genetics or molecular biology. His observation was of the greatest relevance because, as I shall explain in the following pages, these specific proteins have never been discovered and that produces a major contradiction in genetic determinism.

As for me, I went one step further and one thing led to another, ending in my formulating a new theory. During our conversations, I explained to him the main principles which I also described in the first article I had published at that time. Jean Tavlitzki could have prevented me from following this path. He had enough influence over me to do so and I think I would have listened to him if he had tried. He did not do so, however. He did not prevent me from thinking and I am deeply grateful to him for that.

Since this "germinal scene", if I may call it that, I have traversed unforeseen fields such as the philosophy of biology, published several articles to complete my theory, and undertaken various research studies, to arrive now, at writing this book.

In it I explain the theory of cellular Darwinism, also called ontophylogenesis. It is the extension of natural selection, taking place inside the organism among the cell populations of which it is constituted. It ends with evolution and ontogenesis merging into a single phenomenon. Its application models help support it and show that it also emerges onto a concrete experimental research programme.

I have already formulated my theory in my earlier articles, but now I explain how this general theoretical context breaks with both genetic determinism and self-organisation, and how it goes beyond their contradictions. Indeed, since the publication of genome sequencing, many researchers, probably disappointed by its results, are turning to theories of self-organisation, thinking they may find an alternative there to genetic determinism. I have therefore devoted long sections to analysing the foundations of holism and self-organisation in which I show that these theories are only superficially different from genetic determinism, and that they are not a valid alternative.

Finally the story speeds up. Many data published in the last ten years support ontophylogenesis and give it a new dimension. The non-specificity of proteins is nowadays documented to the point where it has led to the foundations of molecular biology being challenged by molecular biologists themselves. Probabilistic gene expression has also become an unquestionable phenomenon. I have undertaken computer simulations which demonstrate the relevance of ontophylogenesis, as well as analyses of its epistemological aspects. All these studies are new material that I have incorporated into this work.

This book therefore mainly concerns biology, but also touches on philosophy and history. Although not a book aimed at the general public, it is addressed to a broad audience extending beyond the circle of specialists. I have avoided terminology which is too technical as much as possible. I have not always avoided redundancy in my explanations where they help comprehension and I have frequently used explicit expressions even though they may make the style more ponderous. I have provided a glossary to help the non-specialist reader and invite him to refer to it as often as necessary. He might even do well to begin by glancing at it. As for the bibliography of the subjects tackled, it consists of a long list. I have more often than not contented myself with referring to the most significant examples or syntheses. Certain chapters also include many quotations. Looking back at these historical texts on which biology is founded is essential to escape from the confusion which reigns in current debates. Reading them requires a little extra effort on the part of the reader.

Acknowledgements

I have been developing my work for five years at the Centre Cavaillès, an ideal location where it is possible to undertake pluridisciplinary research. Michel Morange and Robert Nardone have read my manuscript. Through their remarks and advice I have been able to improve it greatly. I have also had long discussions with Jean-Marc Lelièvre, whose constant support has been of valuable assistance. Hussain Monier and Jérôme Glisse have provided their expertise in creating the figures. In various ways, writing this book has been encouraged by Véronique Anger, Christian Bréchot, Axel Kahn and Denis Noble, while World Scientific Publishing has given me the opportunity of publishing it in English in the best conditions any author could wish.

Je est un autre

~Arthur Rimbaud

Contents

1

Five Arguments for a New Theory of Biological Individuation

Biological theories have been propounded since ancient times. In an attempt to grasp the nature of the species and the individual, and in general, the genesis of these two aspects has been considered as distinct phenomena. This is the reason why the evolution of species and the development of organisms is explained by two different theories: natural selection and genetic programming. This separation presents a recurring problem, as these two processes are in fact closely interwoven one with the other. In concrete terms, the species evolves through the reproduction of individuals that succeed each other. There has to be, therefore, a point where the theory of evolution and the theory of embryonic development meet. In the 20th century, this union occurred through what has become known as 'evolutionary synthesis'. Evolution of the species is considered to arise from transformation due to mutation of the genetic programmes coded in the DNA. This field of research is now called "evo-devo". While, logically, the two processes can be linked to one another with this theory, the cost is considerable. New problems arise bound very closely with genetic determinism, in which the theory ends. DNA becomes omnipotent. It governs evolution through its mutation, and controls the genesis of organisms through the genetic information it contains. Ever since we developed the ability to sequence genomes, the difficulty of holding such a view has been confirmed.

Firstly, there are considerably fewer differences between the genomes of organisms, including those that are phylogenetically distant, than were foreseen. It is therefore difficult to explain evolution by the addition of DNA point mutations. Secondly, it has not been possible to decode the genetic programmes that are supposed to control embryonic development by reading these genomes. There are far fewer genes than seem necessary to explain all the functions performed by an organism. As a result of these limits to genetic determinism, we are now seeing a real change of paradigm, with the emergence of systems biology.

Instead of focusing our understanding of organisms on their DNA, we are trying to see and understand them as systems. We are seeking, in this new context, to find the balance between the influences arising from the various levels, which include the DNA, the networks of proteins, the cell tissues, the organism and the environment. This post-genomic biology requires enormous use of biocomputing to integrate the huge quantities of data collected by large-scale transcriptome and proteome analysis. The aim of these programmes is to identify all the RNAs and proteins in a cell in order to establish a map of the interactions they have with each other in the form of networks. It is thus hoped to arrive at a complete description of how a cell functions. However, scientific progress does not result simply from accumulating data. The observations made depend just as much on the theories which guide the research as on the reverse. Systems biology will not succeed in going beyond the contradictions of evolutionary synthesis unless it also resolves the original problem concerning separating evolution from embryogenesis. To do this, a new conceptual framework needs to be developed.

Ontophylogenesis (or cellular Darwinism) resolves this problem and provides a conceptual context in which DNA is not omnipotent. It breaks with traditional theories by considering embryonic development and evolution as a single process. It consists of applying Darwinism to the interior of organisms, no longer just to the DNA but also to how a cell functions as well. It thus leads to a general conception in which the question of biological individuation can be tackled from a new angle. It is this theory which is the subject of this book, in the course of which various extensions of it will gradually be discussed.

The concept of probability will first of all be analysed in order to understand the difference between determinism and probabilism. This prior clarification is necessary in order to grasp what the intrinsically probabilistic character of cellular Darwinism involves, and to differentiate it from the theories of genetics and self-organisation. The latter use the concept of noise or fluctuation but are fundamentally deterministic theories (chapter 2). The principles of genetic determinism will also be studied in detail. We shall see that they are incompatible with recent experimental data because the molecular order that they imply for explaining biological organisation does not exist (chapters 3 and 4). We shall then examine the variants of holism, theories which assert that order, instead of originating from the molecular level as in genetic determinism, originates from higher levels of organisation. The analysis will show that they are not valid alternatives, as they rely on the idea of a creative nature and a return to animism, which are purely and simply a negation of scientific rationality (chapter 5). We will then discuss ontophylogenesis, which differs from reductionism and holism in that it does not presuppose origin in biological organisation, whether concealed at molecular level or at higher levels of organisation. As a result, ontogenesis can really be considered as a process and not as the expression of a static order. The experimental data supporting it have been accumulating for more than forty years. They show that gene expression is a probabilistic phenomenon and that there exist mechanisms exerting selection on cell differentiation. In addition, computer simulations show that cellular Darwinism is in a position to generate reproducible cell structures and that chance can play a positive role in this process (chapter 6). Finally, ontophylogenesis will be placed in a wider historical and philosophical perspective, which will distinguish it as much from Aristotelian (hylemorphic) conceptions, which place the origin of organisation in Form, as from Hippocratic conceptions that place it directly in the material body taken as a whole. This analysis will again show how it differs from genetics and self-organisation which, for their part, remain within these traditional modes of thought (chapter 7).

All these developments will gradually produce support for the five main arguments for cellular Darwinism, which we shall first of all set out rather bluntly in a summarised form.

1.1 Ontophylogenesis

An adult multicellular organism, comprised of numerous differentiated parts, results from the development of an embryo, which itself arises from the multiplication of a germinal cell. During this process, cells which are undifferentiated at the start become specialised and organise themselves into tissues that carry out the functions necessary for life. An organism with the characteristics of a biological species is thus produced. New germinal cells are in turn generated in this organism and the process of embryogenesis is reproduced cyclically. How can this phenomenon be explained? This is a question which is very difficult to answer.

Up until the present time, the functioning of living beings has always been interpreted in line with deterministic theories. For genetics and molecular biology, the organism is inscribed in advance in the genome as a code containing the genetic programme. The cells differentiate according to the instructions in this programme: the genes are activated in sequence during the development and synthesise specific proteins which serve as signals exchanged by the cells. Under the influence of these signals, the cells differentiate for specific purposes, a totally deterministic phenomenon which excludes chance. This theory, which poses serious conceptual problems, is now being refuted by a large number of experimental facts.

Cellular Darwinism renounces the deterministic tradition of embryology and genetics. Cells change state and differentiate because the way they function is intrinsically probabilistic. There is randomness deep inside them, in the way the genes function, where they are supposed to be controlled by the genetic programme. Depending on whether one set or another of these genes is expressed by chance, from all those that make up the genome, the cell acquires certain characteristics that correspond to a particular differentiated state. Interactions between cells play an important role, but they do not involve signals inducing changes of state, as the theory of genetic programming supposes. Rather, they stabilise genetic expression when a viable combination of differentiated cells has been produced by the genes functioning randomly (Kupiec, 1983). The genetic expression is then frozen and the cells can no longer change their state. If a cell does not adapt to its microenvironment through this random process, it ceases to multiply and dies or becomes pathological. The conceptual structure of this model is therefore a mixture of chance and selection, analogous to the theory of natural selection but transposed to the level of cell behaviour.

However, the analogy with Charles Darwin's theory (1809–1882) goes further. According to cellular Darwinism, embryogenesis is a real extension of natural selection within organisms. Ontogenesis and phylogenesis are the two inseparable sides of a single reality produced by a unique process: ontophylogenesis. Organisms develop and evolve at the same time. Both phenomena are the result of a single mechanism (Kupiec, 1986), so for this reason, the usual definitions of the genome and the environment are not apt. Since it functions randomly, not only is the genome not the bearer of a genetic programme of rigid instructions in which the adult organism is inscribed in advance, but the conception that we have of the environment is equally incorrect. It comprises not only an external environment from which the organism is separated by a hermetic barrier, but it continues inside the organism forming the selective microenvironment of the cell, to which the latter must adapt. This conception of the cell microenvironment corresponds to Claude Bernard's 'internal environment' (1813–1878). For him, organs and cells lead an autonomous life in this internal environment (Bernard, 1878). Cellular Darwinism borrows, therefore, both from Darwin's and Bernard's theories, and consists of applying natural selection to the cells which live in the internal environment. A similarly inspired theory was put forward by Wilhelm Roux (1850–1924) in the 19th century, but it was eclipsed by the expansion of genetic conceptions (Roux, 1881).

1.2 Random man

The term 'Darwinism' nowadays no longer refers to the original theory set out by Darwin but to evolutionary synthesis. Species evolve in this context due to random mutations of DNA, which produce advantages for certain individuals in using the environment. They are selected therefore owing to their more rapid multiplication. It is a question of simplification, which eliminates the fundamental aspects of Darwin's thought. His book *On the Origin of Species* (1859) puts forward a theory explaining the transformation of species, but also questions what a species actually is. The word 'origin' must be understood as meaning 'a mechanism generating the species' and not 'chronological origin' in a history of living forms, which is not what Darwin meant. He first of all considered the definition of 'species' and its significance. What he said about this is very surprising and runs counter to common sense. He began by defending a nominalist vision. He asserted that species do not actually exist in nature, but are abstract entities created by the classifier by arbitrarily grouping living forms together, depending on his subjective appreciation of them. Darwin's nominalist position is nowadays totally suppressed or considered as an error of his youth corrected by the adherents of evolutionary synthesis (Mayr, 1993). As we shall see, it is this most revolutionary and most fertile aspect of his thought that contains the germ of a general theory of living beings. In contrast, genetics is not nominalist but is founded on the reality of the species. Evolutionary synthesis is thus forced synthesis between profoundly contradictory elements, which leads to a theoretical and experimental contradiction (Kupiec, 1999).

The special fields of both Bernard and Darwin have found their way into the realms of modern science, but it is unusual for the two to be brought together. Their individual areas of research seem very separate. Bernard's is concerned with the way the organism functions internally and argues for absolute determinism, while Darwin's concerns the organism's relationships with its external environment, and argues a probabilistic theory.[1] Earlier discussion here has already indicated the similarity between the Darwinian environment and Bernard's internal environment. Their convergence however equally concerns their epistemological conceptions. In the same way that Darwin doubted the objective reality of species, theoretical models of physiology were, for Bernard, creations of the mind, and their reality should be considered as subjective. He goes even as far as doubting the reality of physiological functions. So although nowadays it seems extremely surprising to a biologist who does not know the history of his discipline, the father of the modern theory of evolution did not believe in the reality of the species and the father of modern physiology did not believe in the reality of physiological functions! The profound significance of this shared nominalism must be analysed. It indicates radical anti-essentialism which lets us understand the living being while renouncing any finalism. Cellular Darwinism radicalises this Bernard/Darwinian anti-essentialism.

To break, indeed, with the essentialist biology of Aristotle, Descartes (1596–1650) introduced the idea of the 'Animal Machine' taking the clock as a metaphor for the organism. La Mettrie then extended it to the 'Man Machine'.[2] Reducing the living being to a machine means that the physiological processes follow ordinary physical and chemical laws, like the rest of nature. Living material, like inanimate material, is inert in itself. It is the forces that are each exerted on the parts of a living body[3] that make an organism move and endow it with vital characteristics. However, Mechanism

---

[1] Darwin does not actually use the terminology of the theory of probability. As we shall explain in chapter 7, §7.3, his explanation of the origin of hereditary variations was ambiguous. However, it covered what nowadays we call chance variation.

[2] Descartes considered man to be made of two substances, the immaterial mind and the body. The metaphor of the machine only applied to the body. La Mettrie radicalised this position. For him there is only one material substance. What is called the mind only arises from the organs of thought functioning, thus producing the man machine.

[3] Or chemical reactions between molecules.

does not succeed in totally eliminating finalism because a machine is built by its designer according to a plan decided in advance for fulfilling a function. All the parts making it up are adjusted relative to each other to fulfil this overall purpose. Necessity rules supreme. Cellular Darwinism also considers life as an exclusively physical and chemical phenomenon, and from this point of view, it is mechanistic; but, contrary to traditional mechanism, which is fundamentally deterministic, it is based on probabilistic laws. The Man Machine, as far as cellular Darwinism is concerned, is a 'Random Man' and thus escapes totally from finalism and essentialism.

Cellular Darwinism is also different from theories of self-organisation, which postulate that matter is not inert but on the contrary has creative properties producing life. Man, according to these theories, cannot therefore arise from chance.

1.3 The same kind of laws govern biology and physics

In physics, order is subjective because it is related to the level of observation at which the experimenter places himself and to the degree of accuracy that he sets. Macroscopic order at our level of existence arises from microscopic disorder. The behaviour of molecules and atoms when considered individually is intrinsically random but this molecular disorder is insignificant at the macroscopic level. Due to the huge number of particles making up systems, the individual variability of each molecule is negligible compared with the average behaviour of the whole. Erwin Schrödinger (1887–1961) spoke in this respect of the principle of 'order from disorder' which governs physics. In contrast, as regards molecular biology and genetics, biology is supposed to be subject to a principle of 'order from order'. The order is supposed to be real, intrinsic to the living thing and irrespective of the subjectivity of the observer. The macroscopic organisation of living beings is said to be produced from the microscopic order laid down in the chromosomes in the form of genetic information. This theory holds that biological molecules do not collect together according to the probabilistic laws of physics but fit together according to the instructions relating to this information. There would therefore seem to be a difference in kind between physics and biology. In physics, order would seem to be epistemological[4] whereas it is supposed to be ontological[5] in biology. This analysis by Schrödinger (1944) is the basis for the theory of genetic programming. It has been dominating molecular biology since it started but its historical and philosophical roots are a great deal older. Genetic information is equivalent to the formal cause, or to the soul, in Aristotle's philosophy. It is an order principle which determines an invariable organisation of living beings corresponding to the species. This analogy between genetics and Aristotle's system has already been probed by the founders of molecular biology and evolutionary synthesis (Delbrück, 1971; Mayr, 1982; Mauron, 2002; Vinci and Robert, 2005) without their considering it a problem: the fact that biology uses Aristotelian concepts would only go on to show the relevance of his system. In actual fact, this theoretical structure induces contradictions which undermine the development of the molecular biology research programme.

Indeed, this Aristotelian conception of molecular biology is nowadays invalidated by the most recent observations. Contrary to what it predicts, there is very great molecular disorder in biological systems. Gene expression and interactions between proteins are not rigidly determined, but rather the reverse — they have a fundamentally probabilistic character. Cellular Darwinism goes beyond this contradiction because it takes physical and chemical probabilistic laws fully on board. The behaviour of proteins is subjected to Brownian motion and the laws of diffusion. It does not therefore, as does genetics, introduce a difference in kind between physics and biology. In this respect it is different again from theories of self-organisation which, like molecular biology, consider order to be real.

---

[4] i.e. in the knowing subject.

[5] i.e. constituting that which is real, inherent in the world, irrespective of the knowing subject.

1.4 The first principle of biology

The idea that philosophy no longer has very much to contribute to science is very widespread. Just as science is supposed to be capable on its own of providing us with access to real knowledge, sure of its truth owing to experimental method, philosophy is supposed only to be metaphysical speculation, of absolutely no use for scientific research. It could only be used at best to study the methodology and development of science. Seeing it this way is wrong. In all the sciences there are entities or first principles which serve as starting points. These principles are not demonstrated, but are stated *a priori* as constituting the reality. They are not intangible. For example, Newton's physics is based on three-dimensional space and absolute time. Yet this prime structure of the universe was abandoned by Einstein, which led him to work out the physics of relativity. First principles arise from ontology which is an area on the limits between science and philosophy. The choice of first principles is very important because they determine the nature of scientific theories which are constructed from them.

In biology, the question of first principles does not seem to present a problem. It can be formulated thus: What are the primordial entities of the living world? The answer seems to be obvious: when we look at the living world, we can immediately pick out individuals managing on their own, and if we compare them, we observe subsets among them of identical beings. We can thus identify an initial entity, the individual organism, and the species which is coextensive to it.[6] We do not doubt for a moment that these two entities really exist in nature, irrespective of any subjective divisions we make to pick them out, or theoretical suppositions that we apply. A genealogical line is then conceived as a succession of identical ontogeneses but each with its independent individual reality. This seems simple and natural and has always tended to dominate biology. Yet there is another conception, and this is the one that certain classifiers and evolutionists have tended to adopt. It consists of

[6] Since a species is a set of identical individuals.

extending one's view beyond the individual to see the genealogical line as the prime entity, and no longer the organism. It is possible, indeed, to be interested in the first instance in the relationships between beings that resemble each other. The idea that then appears is that of the relationship which unites them, and is based on the material continuity of living beings occurring through the transmission of a germinal cell and its hereditary material. Earlier we recalled Darwin's nominalism. He did not cling to this negative position. Through this nominalism he rejected the essentialist definition of the species, but in its place he substituted an evolutionary definition. In his eyes, a species is a genealogical line for a group of organisms that have the same common ancestor. In such a concept, the genealogical link becomes the first principle and the organism a secondary entity produced by the process creating that link, i.e. the evolutionary process itself. The organism is an entity which has no existence except as an instant in the continuous process of reproducing organisms. This genealogical idea of the living being is implicit in Darwin and explicitly stated by Bernard (Bernard, 1878).

However, it has nowadays disappeared from contemporary physiology, and biology must confront another contradiction. Since the dawn of genetics, it has been dominated by the point of view which considers the individual organism to be a first principle, whereas living beings are historical productions, the explanation of which requires a genealogical design. Neo-Darwinian synthesis has attempted to resolve this problem but has not managed to do so, as it continues to consider ontogenesis and phylogenesis as arising from two distinct processes. Ontophylogenesis, on the other hand, removes this contradiction because it allows effective synthesis of the two points of view by combining embryogenesis and evolution in a single process.

1.5 Man lost in the Amazonian forest

Another very widely held opinion consists in believing that the difficulties encountered in biology arise from the complexity of the living being. This complexity is supposed to be related to its hierarchical organisation of one level above another: molecules, cells, tissues, organs, organisms and ecosystems. Corresponding to this structural hierarchy is said to be a hierarchy of controls leading to very complicated networks of multiple interactions between components, that cannot be described by simple laws. This hierarchical organisation seems obvious but raises the same question as that concerning species: Is it a first principle? Is it ontologically real? As concerns genetics and the theories of self-organisation the answer is positive: it would seem to be a structure helping constitute the living being. Each level seems to have properties determining how organisms function. Due to this ontological similarity, genetics and self-organisation are confronted with the same pitfalls. Self-organisation, which sets itself up as an alternative, is no more appropriate than genetics, and leads to the same contradictions.

Hierarchical organisation is not, on the other hand, a first principle for cellular Darwinism. We find it difficult to accept this idea because there is a particular epistemological obstacle to biology. This lies not in any intrinsic complexity of living beings but in the extreme difficulty we have in going beyond essentialism in our relationship with them. We always want to endow them with characteristics which differentiate them from the rest of nature. These characteristics are intrinsic, either those coded in the genetic information, or emerging and creative characteristics postulated by self-organisation.

We can understand this difficulty better using an analogy. Everyone knows the allegory of Plato's cave. Here, the situation is different. The man is not a prisoner in a cave but is lost in the Amazonian forest. He has no idea of the geoclimatic context of where he is and can never see the Amazon, the existence of which is unknown to him. Before his eyes he has this extraordinary accumulation of vegetation comprised of all sorts of plants, large and small, which are intertwined in every direction. This forest, with its innumerable details, appears to him to be extraordinarily complex and he thinks that the explanation for it must be similarly complex: he seeks a meaning and reason for each detail. Why, for example, is this particular plant exactly in this specific place and why are its branches intertwined with those of that other plant? In this quest for meaning, he succeeds in classifying several types of plant depending on their size, and recognises thus several levels populated with living things that maintain specific relationships which seem to support this complexity: the phenomena connected with small plants that survive close to the ground, with those that occur higher towards the tops of the tall trees, and with those that are at a height in between. This structure seems to him to be inherent to the forest and even to account for it. In fact, each of these levels seems to have its own properties in terms of light, temperature, humidity and sensitivity to wind. Nevertheless, if he were to see the Amazon, he would probably understand that this apparent complexity and apparent hierarchical organisation has a simple explanation related to the abundance of water in this region, which favours the growth of luxuriant vegetation. He would also understand that the multitude of little details that make up the forest are the result of the vagaries produced during this growth, which have neither an explanation nor any particular meaning. As for the levels of organisation, they are not a constitutive structure inherent to the forest but the result of plant growth in the conditions where they are produced. If these conditions change (less water, a different temperature etc.), the structure of the forest would also change, because it is not constitutive but the result of a process conditioned by the structure of the environment.

When we analyse living matter, we are in a situation similar to this man lost in the forest. In the same way that he is unable to see the Amazon, we also have a blind spot. The idea of a natural hierarchy is intimately linked with essentialism which assumes a hierarchy of forms or essences that give structure to the world. This hierarchy ends with Man whose existence has been endowed with a meaning that emanates spontaneously from his nature. It makes him the centre of and ultimate project of Creation. We are incapable therefore of renouncing this because that would mean abandoning our privileged position and recognising in ourselves Random Man, with the loss of meaning that it implies. This prospect is a threat to our integrity and we seek to avoid it at all costs.

2

What is a Probabilistic Process?

SUMMARY. Probability, according to the subjective conception of it, measures the degree of belief we have that certain events will occur; according to the objective conception, it is the result of intrinsic properties of the phenomena which are then produced with a certain degree of frequency. Probability is not, however, incompatible with either causality or reproducibility. In the mathematical theory of probabilities, there is no difference in nature between determinism and probabilism; determinism is just the limiting case of probabilism when the probability of an event is equal to 1. In contrast, for essentialism, the specific and the accidental belong to two levels of reality that are qualitatively different. Countless errors related to using probability arise from confusing it with the essentialist concept of accident. If these two notions are differentiated, the gap that separates an intrinsically probabilistic theory, such as Darwinism, from a deterministic theory with noise, such as self-organisation, can be understood. Darwinism fully implies the modern meaning of probability and not the essentialist notion of accident. As far as it is concerned, order is relative, depending on the relationship of the organism to the environment. Self-organisation reduces what is random to the level of accident. In this case, order is absolute. It is inherent in the organism, and depends on the specific relationships between its components.

Before envisaging a theory of biological organisation based on probabilistic laws, it is necessary to accurately define the concept of a probabilistic process. This is necessary because, while calculation of probabilities pervades all scientific fields, the nature of the process is still very poorly understood. There are frequently confusions related to using it, including in scientific papers. They come from modes of essentialist thought which are contrary to modern scientific practice. They consist in confusing notions of accident, contingency and noise with the notion of probability.

2.1 There is no qualitative difference between determinism and probabilism

We all have some idea of what differentiates a deterministic system from a probabilistic system. We think we know with certainty that certain phenomena are invariably repeated when the same cause is activated. For example, if I let go of this pen I am holding in my hand I have absolutely no doubt that it will fall, and that that is what will happen every time I let it go in the future. This is an event subject to determinism and I can say that the system governed by gravity, including myself holding the pen, and the ground, is a deterministic system. We also know that the type of situation that I have just described is not universal. There are other phenomena for which there is no sole certain effect but which may have several effects resulting from the same cause. The standard example of this is the game of heads or tails. When I toss the coin there are two possible results and I cannot say with certainty whether the result will be heads or tails. This system is probabilistic. In addition, I do not need to be a great mathematician to know that each of these two events has only a 50% chance (1 in 2 probability) of occurring. I also know that probability is always between 0 and 1 and that the frequency of an event occurring when it is repeated several times is all the greater the closer its probability is to 1. When the probability is equal to 1, the event is always reproduced. The difference between a deterministic system and a probabilistic system is therefore connected with the value of probability. This can be expressed quite simply: a deterministic system is a system in which the probability of events occurring there is equal to 1, whereas in a probabilistic system several events are possible, each having a probability between 0 and 1.

This apparent simplicity conceals enormous difficulties. I said for example, "If I let go of this pen I am holding in my hand, I have absolutely no doubt that it will fall and that that is what will happen every time I let it go in the future." Such an affirmation poses the problem of induction raised by David Hume (1711–1776). What logic allows us to be certain that an event that is produced in certain conditions will necessarily be reproduced in the future? The sun rose today as it does every day but that does not prove that it will rise tomorrow. Yet science is constructed based on this type of inductive reasoning whenever we formulate universal laws from particular experiences. Indeed, we seem to acquire such a high degree of belief in the fact that the experience will repeat itself that it attains a degree of absolute certainty. Probability would therefore seem to be a measure of the degree of belief we have in certain events occurring, in relation to our level of knowledge or ignorance of these events. This is the subjective conception of probability first proposed by Hume (1739) and developed subsequently by numerous scientific philosophers in different forms.

However, there is another conception of probability, known as objective or frequency probability. We are aware that certain phenomena are produced with a certain constant frequency when the same event is repeated a very great number of times, for example, the coin falls tails up in 50% of cases. It really seems that it demonstrates an intrinsic property which arises from the physical structure of the coin. According to the objective conception, probability would seem to reflect this intrinsic property of the coin. It does not here reflect our ignorance but is an objective property of the world. Several philosophers have tried to present theories for this objective existence of chance. For many authors, the two aspects of probability, subjective and objective, have always coexisted (Hacking, 1975; Martin, in press). Debates on this question are fraught with difficulties and are not the subject of this book. We shall simply state that the objective existence of chance today seems to have been affirmed by quantum physics and that Karl Popper (1902–1994) developed a totally probabilistic philosophy, according to which randomness is an underlying factor in the world. He speaks of a universe of propensities (Popper, 1992) and defines probability as a disposition or propensity of the phenomena themselves. Such a philosophy is in complete opposition to essentialism which maintains that things are determined by their essences (see this chapter, §2.2.3). In developing the subject later, we shall merely define deterministic and probabilistic systems in the usual way depending on whether the probability of the events is equal to 1 (deterministic) or between 0 and 1 (probabilistic). It is important to note that in this framework defined by the theory of probabilities, there is unity among natural phenomena, because there is no qualitative difference between determinism and probabilism. Determinism is the limiting case for probabilism when probability is equal to 1.

2.2 Errors related to using probability

Opinions of the area covered by the idea of chance are often somewhat vague. The mathematical theory of probability appeared in the 17th century but events subject to chance had already been described through other concepts before that. Those earlier ways of looking at things have not totally disappeared, and still cause ambiguities which are the source of errors that need to be eliminated before analysing biological problems.

2.2.1 *Probability does not deny causality*

One of these errors is to believe that a probabilistic phenomenon, often called indeterministic, does not have a cause. The example of the game of heads or tails shows that this is false. There is indeed a cause, tossing the coin, with two possible results. The word 'cause' means the conditions, in the widest sense, of the random event. A genetic mutation is, for example, a random event. The frequency of genes mutating is determined by the structure of the DNA and the presence of mutagenic chemical agents. It varies depending on the conditions in which the chromosomes find themselves. The probability of a phenomenon always depends on the physical and chemical conditions for it to occur, and in this sense, one cannot speak of indeterminism here. Probability does not deny the causality of phenomena: it just opens up to plurality the relationship between the cause and effect, which is one-to-one in the context of determinism.

2.2.2 *Probability is not incompatible with reproducibility*

Another very widespread error which must be avoided in a discussion dealing with the mechanisms of embryogenesis consists in believing that a probabilistic phenomenon is not reproducible because it involves chance. On the contrary, the concept of probability expresses the existence of order and reproducibility where there was thought to be none, before it was conceived.[7] When chance is rationally mastered by mathematics, predictions can be made with a very great degree of accuracy by calculating probability. This is commonly done nowadays in modelling numerous natural or economic processes.

We have already seen that the probability of an event is seen in the frequency of its occurrence being stable when it is repeated a very great number of times. To be precise, the definition of the probability of an event X occurring is its frequency as the number of experiments performed approaches infinity. In practice, if we repeat an experiment involving random events a very great number of times, each time we perform it the events occur with constant frequency, ignoring minute negligible deviations. For example, if I play heads or tails, the frequency of each of these events will always be 50%. Probability thus expresses a stable structure of the world which is not manifested by individual events but by populations of events which are repeated a great many times. Unlike common sense, probability expresses reproducibility where it is not immediately obvious. It indicates that the order of the world is not

[7] The calculation of probabilities was invented in the 17th century by Pascal to solve problems connected with games of chance.

absolute but relative to the individual or population level in which one is situated.

A random phenomenon is therefore statistically reproducible for a population of events. This reproducibility is described using two parameters, mean and variance. Everyone is familiar with the concept of a mean. Variance is a measure of the variability of a variable compared with its mean.

Figure 1 shows examples of distribution of a random variable with different variances. In such distributions, if the variance is very small, a phenomenon can seem to behave like a deterministic phenomenon even though it is probabilistic. Indeed, each time it is produced, the results, which are very close to the mean, seem identical. This is all the more true when the law of large numbers is applied, if it is a phenomenon itself composed of a very great number of random events. The variance of a phenomenon composed of a series of random events is in fact reduced as the number of events

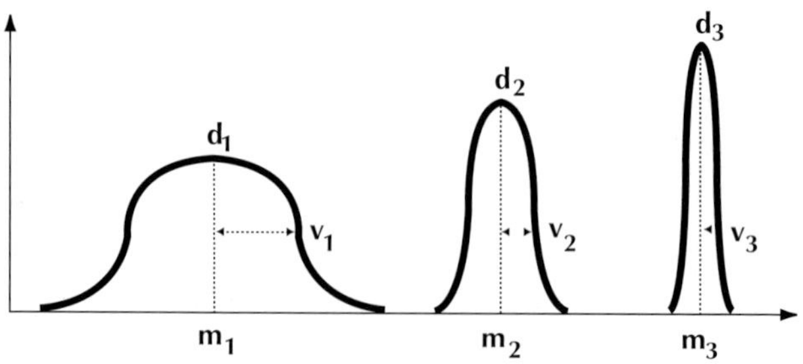

FIGURE 1. **A random phenomenon is statistically reproducible.** In a population, a random variable describing a probabilistic phenomenon is distributed according to a statistical distribution shown by the mean m of the variable in the population and its variance v which shows its variability around this mean. d_1, d_2 and d_3 are statistical distributions of this type and have different means and variances. Two distributions may nevertheless have the same mean but different variances. This would be the case with two classes where the mean of the grades obtained by the pupils for their maths homework is 10. In one of the classes however, the grades range from 2 to 18 while in the other they only range from 9 to 11. The variance in the first class is greater than in the second. A probabilistic phenomenon with a small variance, such as that shown by d_3, may seem deterministic because each time it occurs its variance is close to the same mean value.

increases. This law can be understood easily without doing complicated calculations.

Let us return to our game of heads or tails. We begin by doing series of ten tosses. The theoretical frequency of heads or tails is 50%, but in practice we would not obtain that for each series of ten tosses. In certain cases we may obtain, for example, seven tails and three heads, which correspond to frequencies of 70% and 30%. If we then do series of a million tosses, it is in this case impossible to obtain 700 000 tails and 300 000 heads complying with the earlier frequencies of 70% and 30%. Owing to the law of large numbers, the frequencies observed will be very close to 50%. The number of deviations compared with this theoretical frequency will even be so low that it can be ignored. When we repeat several series of a million tosses, the result will then appear constant and if we only have access to this overall global result without knowing the details of the experiment, random tossing for heads or tails, we may think that the phenomenon is deterministic whereas it is probabilistic.

We will come back to this crucial problem in the next chapter. Schrödinger drew very important consequences from this law of large numbers which have profoundly marked molecular biology.

2.2.3 *Probability, accident and contingency are not synonymous*

The word 'accidental' is often used in place of probabilistic. This is an approximation which leads to misinterpretations. The major contribution of calculating probabilities is to introduce rational control of events subject to chance and include them in a scientific analysis. In contrast, the concept of accident is of pre-scientific origin and using it constitutes regression in reintroducing the irrational.

Philosophers and scientists have, since ancient times, recognised the difference between deterministic and probabilistic phenomena, but up to the 17th century the difference between the two was believed to be qualitative. Only deterministic phenomena were considered accessible to science, which consisted, in the essentialism inherited from Aristotle[8] (384–322 BC), in trying to define what was constant either in living or non-living things. The characteristics thus identified were called specific characteristics or differences. For example, man was defined as a reasonable animal. The specific difference of 'reasonableness' set him apart from all other animals and at the same time defined him. All the specific characteristics of a thing corresponded to its essence (or nature) and allowed it to be defined as a species. It must be understood, however, that it was not a question of simply classifying objects as we do today. The essence acted like an inherent active ingredient which determined every aspect of a thing because its aim was precisely to bring about the final cause. In the case of a living being, it guided its embryonic development and its physiology. In the case of a physical object, it caused its motion towards its natural place.[9] In this essentialist framework, knowledge was therefore the knowledge of essences. Individual characteristics which are not constant from one being to another in the same species were qualified as accidental. For example, among humans some men are tall while others are small, but that does not change the fact that these individuals are human in nature. However, these innumerable accidental differences, which occur in addition to specific differences, were not able to be the subject of any kind of scientific knowledge owing to their erratic character and because they were not part of the nature of beings. There was therefore a qualitative difference between the specific (the natural or essential) and the accidental: they belonged to two separate orders of reality.

As we have seen earlier, such a difference does not exist between a deterministic system and a probabilistic system defined by the mathematical theory of probability. This difference assumes an essentialist mode of thought based on notions of specificity and

[8] The prevailing idea of the time.

[9] The natural place was the place considered to correspond to the nature of the object where it achieved the state of rest. A heavy object was thus set to move downwards and a light object upwards.

finality. An accident is something that escapes the final cause.[10] To understand how essentialism differs from the point of view of modern science, let us consider an example given by Aristotle. If it rains often in winter, the essentialist will say that that complies with the nature of winter. On the other hand, if one day it is hot, he will say that that is accidental, i.e. not in the nature of winter (Aristotle, Physics, II, 8, 199a). The fact that it can be hot one day in winter is not nowadays considered contrary to a certain nature of that season. We know that this event may occur with some degree of probability, and that, because there is no essence (nature) of winter, this probability will vary depending on the particular geophysical parameters of the place in question. It will be different in Norway from Morocco, at sea level from the top of a mountain, in a polluted town from the countryside etc. Ultimately the probability of it being hot near the equator in winter may be higher than of it being hot at the North Pole in summer.

The word 'contingent' is also used to describe a phenomenon subject to chance, but its precise meaning is different. Contingent means something that is not strictly necessary, which may or may not be. All probabilistic events are indeed contingent but the reverse is not true. Some unnecessary events may not be probabilistic, particularly any action supposed to depend on divine will. Since the power of God is supposed to be absolute and He has complete freedom, His actions contain no element of necessity. Contingency is sometimes used with this meaning. It has obviously nothing in common with the scientific concept of probability.

---

[10] Ian Hacking (1975) wondered why the calculation of probabilities arose in the 17th century. According to our analysis, it seems that it was necessary to destroy the specific/accidental duality to be able to bring random events into the domain of rational knowledge, i.e. the essentialist mode of thought had to be abandoned. This is exactly what happened in the 17th century, marking the height of the scientific revolution.

2.2.4 *Probability is not noise*

The concept of noise is equally problematical. It is very frequently confused with probability whereas its meaning is different and closer to accident. Certain phenomena can be scrupulously described and predicted by deterministic scientific laws, but when an experiment is performed there is always a slight discrepancy between the prediction and the observation. This discrepancy is called experimental background noise. It is due to the fact that one can never perform a perfect experiment: there is always inaccuracy in the measurement, due to the apparatus used or some disturbance to the experiment from an external factor. The experimental approach consists of attempting to isolate the phenomena by getting as close as possible to perfection, but background noise can never be totally eliminated. Measurement is always within the framework of a margin of statistical error. This noise relative to the theoretical result varies randomly from one experiment to another but, if the phenomenon studied is deterministic, it does not nevertheless transform it into a probabilistic phenomenon. Whatever its quantitative importance, there is always fluctuation which is added to the constant result.

A phenomenon subject to noise therefore has to be differentiated from an intrinsically probabilistic phenomenon. These ideas are at the heart of fundamental debates in biology. Recent experimental results suggest that gene expression is a probabilistic phenomenon. These results are however interpreted as noise by numerous authors who think, despite everything, that cell function is a fundamentally deterministic phenomenon and that there is no reason to challenge the theory of genetic programming. On the contrary, if gene expression and cell function are intrinsically probabilistic, that leads to a major upset in understanding biological mechanisms and calls into question genetic determinism (see the following chapters).

To understand the difference between a deterministic mechanism with noise and an intrinsically random mechanism we might consider the example of a car which has to travel between two towns. For a deterministic mechanism with noise, a constant speed is set that the car must maintain throughout the whole journey. If the journey is travelled several times the car will cover the distance in a constant time related to the speed set, affected by some greater or lesser statistical variation depending on the driver, the car and the conditions in which the journey is made (Fig. 2A). For example, on a hill or when there is a violent gust of wind the car may adjust its speed more or less easily depending on how powerful it is. The car's behaviour is deterministic but there are internal disturbances to the system, related to its physical characteristics, and external ones due

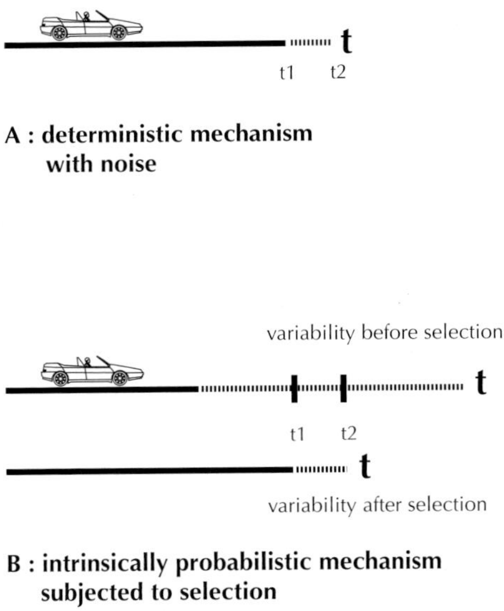

FIGURE 2. The difference between deterministic, probabilistic, and selective mechanisms. The car travels the distance in a time t which varies between t_1 and t_2. **A:** Even if the behaviour were deterministic, there is always inevitable variability (dotted line) in this travel time which is added to a constant time (solid line). **B:** In general, if the car adopts intrinsically probabilistic behaviour without a precisely set constant speed, the variability in its time to travel the distance will be greater and will comply with a statistically different law. If this intrinsically probabilistic behaviour is however, subjected to forced selection, the remaining variability may be comparable to that of a deterministic mechanism. In this case, the two types of mechanisms are difficult to distinguish, except by performing a more detailed analysis.

to the environment. The phenomenon is deterministic with noise. For genetics, ontogenesis is this type of phenomenon. The generation of a phenotype from a genotype is indeed considered a deterministic phenomenon affected to some extent by noise. This noise, which arises from variations in the internal molecular mechanisms and interaction with the environment, disturbs the deterministic action of the genes. It causes small differences between individuals which increase the diversity of phenotypes. There are countless phenomena of this kind in nature. When noise has negative effects it must be limited or compensated in order not to destroy the phenomenon in question, but it can also have positive effects by permitting a system to change state. The standard example of this is the reaction-diffusion mechanism analysed by Alan Turing (1912–1954).

In general, if a system in equilibrium is subjected to fluctuation, it will return after a certain lapse of time to its initial state of equilibrium or it will oscillate around this equilibrium with minor fluctuations. A very simple case of this is that of local fluctuation in the concentration of a solute in a solvent. On average, the concentration always remains the same. Turing demonstrated that in certain cases it may be otherwise. If the system is more complex and the relationships between its constituents permit, instead of returning to the initial state after fluctuation, it will evolve towards a different equilibrium. Turing suggested that such a reaction-diffusion mechanism seemed to be the basis of morphogenesis in living beings[11] (Turing, 1952).

The principle of this mechanism is easy to understand. Take two biochemical compounds, an activator (Ac) and an inhibitor (I). Ac activates its own synthesis and also that of I. I inhibits the synthesis of Ac. Ac is diffused a great deal more slowly than I. Initially, they are present along the axis of an embryo at a constant concentration. This system is in equilibrium because in all respects the activator effect of the molecules of Ac is compensated by the inhibitor effect of the molecules of I (Fig. 3A). If, at a point k,

[11] Numerous analogous systems have been described today. They are known as metastable (or multi-stationary) systems.

fluctuation caused by Brownian agitation of the molecules increases the concentration of *Ac*, its activator effect at this point becomes greater than the inhibitor effect of *I*. If, in addition, this fluctuation is sufficiently great, the system will not return to its initial state. The autoactivator effect of *Ac* on its own synthesis is increased at the same time as activation of the synthesis of *I*. The result is an

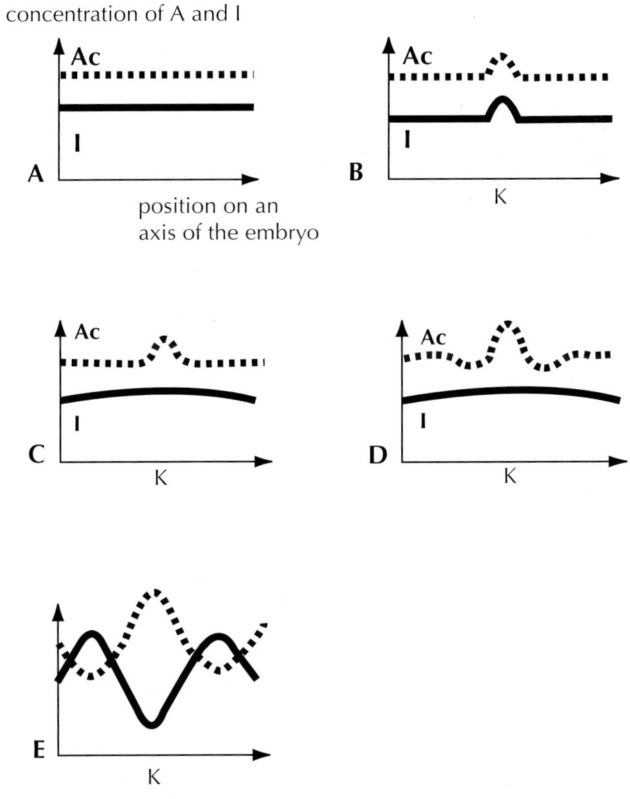

FIGURE 3. The reaction-diffusion mechanism. A: The activator *Ac* (dotted line) and the inhibitor *I* (solid line) are in equilibrium, in a homogeneous concentration in an embryo. **B:** This system is disturbed at a point *k*. The result is local fluctuation of the concentrations of *Ac* and *I*. **C:** The disturbance of *I* is propagated more quickly than that of *Ac*. **D:** Consequently, at certain points the inhibitor effect of *I* becomes greater than the activator effect of *Ac*. This results in reduction in the concentration of *Ac*. **E:** This effect increases until the system reaches a new heterogeneous equilibrium.

increase in the concentration of Ac and I at point k (Fig. 3B). But, Ac and I diffuse in space: because I diffuses more quickly than Ac, its concentration increases more rapidly to the right and left of k (Fig. 3C). The inhibitor effect of I on the synthesis of Ac therefore comes to dominate in these regions and, consequently, the concentration of Ac decreases while I continues to accumulate there (Fig. 3D). This process continues until the system arrives at a new state of equilibrium characterised by alternating peaks of concentrations of Ac and I (Fig. 3E). Therefore in the end the system moves from a homogeneous to a heterogeneous state. According to Turing, this concentration heterogeneity of Ac and I is the starting point for morphological differentiation of the embryo, owing to the local differences it implies in the properties of the system.

In this metastable system there is certainly a random event, the fluctuation, which triggers the reaction-diffusion mechanism, but, once it has been produced, the dynamics of the system are totally deterministic. They depend on the specific relationships between the constituents, Ac and I, and are defined by a system of equations without a random variable.

The concept of noise is also the basis for the theories of self-organisation. These theories state that, as in reaction-diffusion mechanisms, if the disturbances affecting a system exceed a certain threshold, instead of returning to its state of equilibrium the system will evolve towards another more complex state known as an attractor state. New characteristics will then emerge which cause qualitative changes. Those favouring these theories have suggested several different mechanisms to account for these emergence phenomena (see chapter 5).

To illustrate now the case of an intrinsically random mechanism, let us return to the example of the car. It has to do the same journey as the one demonstrating determinism with noise, but its speed, instead of being constant, varies in a totally random fashion all the time.[12] If the journey is repeated several times, the car will

[12] The speed is continually determined randomly. In view of technological progress it would be possible to construct such a crazy car.

cover the distance in a time that varies in line with a statistical distribution (Fig. 2B). This is a case of an intrinsically random phenomenon because the behaviour of the car is itself random. The statistical variations are not the result of disturbances to a deterministic mechanism. In addition, selection can be imposed on the intrinsically probabilistic behaviour of the car. Only certain journeys may be chosen, for example those made between two times fixed in advance. That will restrict the distribution of the results and it may even be that the remaining variability will be of the same order of magnitude as that of a deterministic mechanism with noise (Fig. 2B). Nevertheless, in this case the phenomenon remains intrinsically probabilistic and the variability of the result is reduced by the selection applied. This is a phenomenon that in concept is analogous to natural selection, where the diversity of phenotypes produced randomly by mutation is reduced by selection.

As everyone knows, the theory of natural selection is very important in biology. It implies that if there were no selective constraints, all living forms would form but one continuum, in which no form was distinguished from the others as a result of their continuous variation (Fig. 4A). It is solely because of the action of environmental selection that species can be separated one from the other (Fig. 4B). Thus the order they represent is not intrinsic to the organisms but is the result of their relationship to the environment. In addition, it is relative. If the selective constraint changes, other species are selected (Fig. 4C). That explains why, although all species have a common origin, they populate different ecosystems. In contrast, in a deterministic mechanism with noise, such as reaction-diffusion or self-organisation, the order only depends on the specific relationships of the constituents which are intrinsic to the organism. Order is therefore absolute and inherent in the organisms. The random event is integrated into these theories as noise, that is to say, as an accident that triggers the specific reaction-diffusion or emergence phenomenon.

It is important to differentiate deterministic mechanisms with noise clearly from intrinsically probabilistic mechanisms. It is true that superficially they resemble each other in calling on random

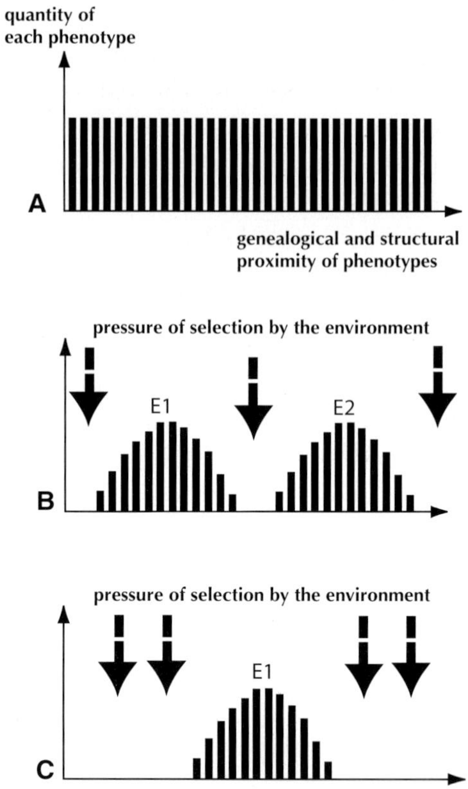

FIGURE 4. The relativity of the order produced by natural selection. Living beings are subject to small variations in each generation. There is therefore a high correlation between their genealogical and structural proximity. **A:** If they were entirely and only subject to this variability, each form would be imperceptibly different from its closest relative. All living beings would form a continuum. No species would be created. **B:** To create a species the action of selection is needed to eliminate or promote certain variants rather than others. **C:** Different variants are selected and different species formed depending on the characteristics of this selection, which change from one environment to another.

events — but they are based on diametrically opposed theoretical contexts.

There is enormous statistical variability in the physiological phenomena which occur inside organisms. This has always been interpreted as resulting from deterministic mechanisms with noise, whether considering genetics or self-organisation. If it were to be found, on the contrary, to be the result of intrinsically random mechanisms submitted to selective constraints, that would constitute a major change in our understanding of the living beings.

3

The Determinism of Molecular Biology

SUMMARY. In non-living physical systems, macroscopic order arises from the probabilistic behaviour of atomic particles, the variance of which is reduced to a negligible level by the law of large numbers. In genetic determinism in contrast, biological organisation arises from intrinsic molecular determinism. This difference in nature between the laws of biology and those of physics reflects a radical ontological separation between these two disciplines. For physics, order is just a subjective approximation, while for molecular biology, it is real and objective. This is why determinism in biology seems insurmountable. Biological molecules escape Brownian motion because they are highly ordered by the genetic information contained in the DNA. This produces specific interactions between molecules, allowing them to self-assemble and for the organism to be constructed through increasingly complex levels of organisation, controlled by the genetic programme. For this theory to be valid, however, it must be subjected to an imperative. Interactions between biological molecules have to be unequivocal, or very few in number, in order to exclude chance.

Since we entered the era of post-genomic biology, it is often claimed that it has become essential for biologists to be interdisciplinary. In fact, it is worthwhile to recall that interdisciplinary collaboration has existed since the early days of molecular biology, which consists of applying physical and chemical methods to biology. Physicists have been at the forefront in this. The most well-known is probably Francis Crick, the co-discoverer of the structure of DNA in 1953.

The influence of other, less famous physicists has nevertheless been even more important.

Molecular biology was initiated by a group of researchers led by Max Delbrück (1906–1981). After working in the field of quantum physics he sought to understand the molecular foundations of heredity. To do this, he studied the multiplication of a bacterial virus and received the Nobel Prize in 1969. Schrödinger, also a physicist, played a major role in forming the concepts of molecular biology. While he is indeed famous above all for his work in quantum physics, he was also very interested in biology, which led to his writing a little theoretical book entitled *What is Life?*, in which he analysed the problems of biology in an original way, basing his thoughts on certain key ideas that had been suggested by Max Delbrück (Schrödinger, 1944). This book had an enormous influence on the founders of molecular biology.

Michel Morange has already presented a detailed study of the history of molecular biology highlighting its techniques and key concepts (Morange, 1994). We are only interested here in briefly reviewing some of these concepts, those that pose a problem for understanding ontogenesis today in the light of the most recent research results (chapter 4). Our starting point is Schrödinger's analysis.

3.1 Order from order

In *What is Life?* Schrödinger considers the origin of order in biological systems and wonders whether the laws of physics can account for it. The answer he produces is negative and he ends his reflection by asserting that there is indeed a difference between the laws of physics and those of biology.

He begins by explaining that "*Physical laws rest on atomic statistics and are therefore only approximate*" (WIL p.10). For, in fact:

> "*We know all atoms to perform all the time a completely disorderly heat motion, which, so to speak, opposes itself to their orderly behaviour and does not allow the events that happen between*

a small number of atoms to enrol themselves according to any recognisable laws. Only in the co-operation of an enormously large number of atoms do statistical laws begin to operate and control the behaviour of these assemblies with an accuracy increasing as the number of atoms involved increases. It is in that way that the events acquire truly orderly features" (WIL p.10).

In other words, order in macroscopic physical systems[13] arises from molecular disorder. Atoms and individual molecules behave randomly. If any order is produced it is solely due to the law of large numbers,[14] which reduces variability to a negligible level in the immense populations of particles making up physical objects. Schrödinger gives several examples of this general principle, the most important of which, for biology, concerns Brownian motion and diffusion.

Owing to thermal agitation, atoms and molecules are continually bumping into each other and moving about randomly, which causes diffusion. This phenomenon can be described, despite the probabilistic character of the movement of each individual atom, by deterministic equations on the macroscopic scale. Indeed, if you put a drop of a coloured product in a glass of water and then analyse the concentration, you will find after a certain time that it is uniform throughout the glass. If the experiment is repeated in identical conditions, the phenomenon of diffusion will be repeated in exactly the same way despite the random movement of each atom of coloured product. This is due to the huge number of atoms involved in this phenomenon. The variance is so small from one experiment to another that in practice we only observe the mean effect of all random individual movements (WIL pp. 14, 15).

The question then arises of whether order in biological systems may proceed from such random dynamics, in which variability would be eliminated in the same way through the law of large numbers. Schrödinger accepts the very strong determinism postulated by the

[13] Those existing at our level of observation.

[14] See previous chapter (Sec. 1.1.2)

founders of genetics. In his opinion, living beings are totally controlled by chromosomes. Although at that time DNA had not yet been identified as the fundamental genetic material of heredity,[15] Delbrück had already suggested that hereditary material might be composed of an aperiodic crystal in which the order of atoms determined the properties of the genes.[16] Schrödinger adopted this hypothesis and suggested that chromosomes contain a virtual code representing the organisms and the way in which they function: "*It is these chromosomes, (...) that contain in some kind of code-script the entire pattern of the individual's future development and of its functioning in the mature state*" (WIL p. 21). In his view, someone able to decipher this code would be able to foresee the organism concerned and fully understand how it functions. Could these organisational properties of chromosomes arise therefore from the random behaviour of the atoms that make them up, as happens in physical systems? Schrödinger asserted that this was impossible because they are not numerous enough, and to consolidate this assertion he relied on a calculation he performed using data available at the time. Obviously his reasoning may seem perfunctory today, but he had no idea of the mechanisms of protein synthesis with which we are now familiar. He estimated that a gene was composed of at most a million atoms, which is very few compared with the size of physical systems, and would not allow the law of large numbers to eliminate variance in the behaviour of these atoms. As Schrödinger put it, "*That number is much too small (...) to entail an orderly and lawful behaviour according to statistical physics*" (WIL p. 30.) Consequently, in contrast to physical systems, order must already be present in biological systems at the molecular level itself and must be responsible for the unique properties of living beings:

> "*... we are here obviously faced with events whose regular and lawful unfolding is guided by a 'mechanism' entirely different from the 'probability mechanism' of physics. (...) Whether we find it aston-*

[15] In general it was thought to involve proteins.

[16] As we know, this hypothesis was confirmed subsequently with the discovery of the genetic code carried by the structure of DNA.

ishing or whether we find it quite plausible that a small but highly organized group of atoms be capable of acting in this manner, the situation is unprecedented: it is unknown anywhere else except in living matter" (WIL p. 79).

"*Two ways of producing orderliness*" have to be distinguished therefore. Whilst physics is subject to a principle of "*order from disorder*", biology is based on a principle of "*order from order*" (WIL p. 80). Biological molecules escape Brownian motion. They are guided by this order principle, i.e. the code contained in the chromosomes that is now called genetic information.

This principle is at the centre of molecular biology and leads to a radical ontological difference between physics and biology. In physics, order is not an objective property of systems, but rather a subjective approximation relative to the observer's level. In biology on the other hand, order is a true property of the organisms, independent of the observer (Fig. 5). This conclusion is of the utmost importance. It may seem a long way from experimental practice but its consequences directly condition the development of biology. It directly influences the way problems that are posed in biology today are dealt with, because it implies that it is subject to insurmountable determinism.

3.2 Stereospecific self-assembly

To understand ontogenesis in the context of genetics and molecular biology, we need to understand how the phenotype is produced from the genotype. The problem comes up against a particular difficulty, however. How can a representation of the organism which is coded in the genome in the form of information be transformed into a real, three-dimensional, phenotypic structure during embryonic development? In other words, how can a virtual organism be materialised and what is the physical process that allows such a transformation?

Schrödinger attempted to answer this question. In the final pages of *What is Life?* he suggested that physical, totally deterministic laws specific to biology must be at work in living beings.

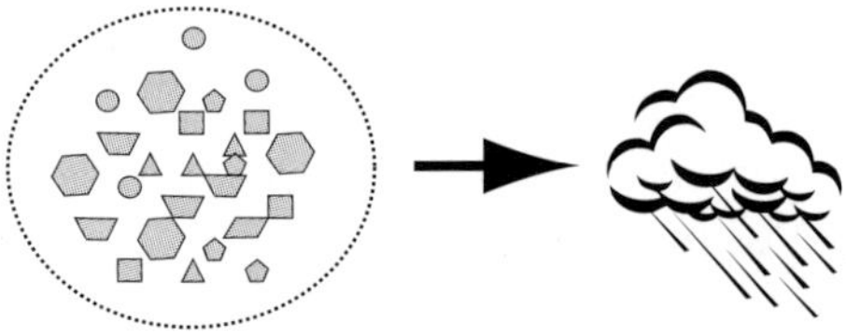

A: In physics: order from molecular disorder
(effect of the law of large numbers)

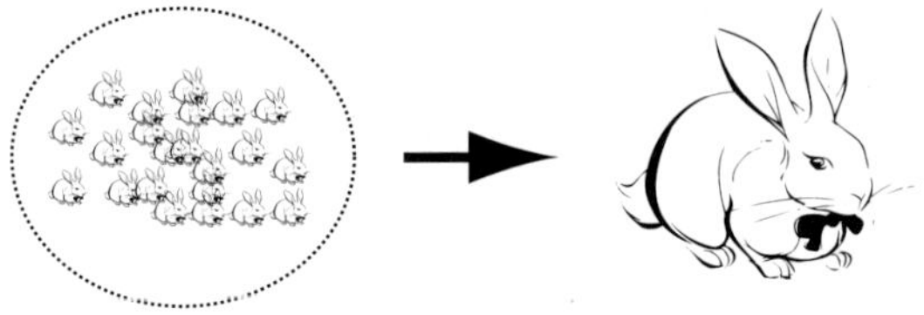

B: In biology: order from molecular order
(effect of genetic information)

FIGURE 5. The ontological difference between physics and biology (according to Schrödinger). A: In physics, order is a subjective approximation. The different molecules are subject to heat motion. The order we see is due to the law of large numbers which reduces the variability of phenomena owing to the huge number of molecules involved and the conditions in which the phenomena occur. **B**: In biology, on the other hand, order is objective. The molecules escape heat motion. They carry information relating to the macroscopic living being (the genetic information).

The principle of order from order must be based on these laws, and it must be due to them that the living being functions with extreme accuracy. Schrödinger made use of Descartes' metaphor for his own ends. He compared the mechanisms of the living being with those of a clock (WIL pp. 81–85). In his view, genes would be like physical matrices from which information would be propagated into the cell by a series of mechanisms as accurate as the transmission of movement inside a clock by cogwheels. It has not been possible with subsequent development of molecular biology to elucidate these physical laws specific to biology. However, the principle of order from order involving a molecular mechanism eliminating chance has been maintained, and the idea of cogwheels working a clock has remained in models for molecular biology with the notion of stereospecificity.

Between 1950 and 1970, two series of work helped formulate a theory explaining the construction of a phenotype from genetic information. In the first instance, protein synthesis was elucidated, with the discovery of the transcription of genes into RNA and the translation of RNAs into proteins. In the very strong genetic determinism which reigned at the time, this led to the 'central dogma of molecular biology' enunciated by Francis Crick (1958), which asserts that genetic information can only be transferred in one direction, from the DNA to the proteins, with no possibility of it being transferred back to the DNA. This therefore prevents the organism from influencing the genome in any way and gives DNA absolute power over biological processes. Secondly, the molecular interactions involved in morphogenesis and the regulation of biological systems were analysed. The existence of a property of stereospecificity[17] was then advanced. The concept arises from the lock and key model suggested by Fischer in the 19th century for defining relationships between an enzyme and its substrate (Fischer, 1894). It has been used in immunology to explain recognition by an antibody of its antigen. Then, as Jacques Monod (1910–1976) explained in *Chance and Necessity* (1970), it has been generalised to all molecular interactions, including those involved in morphogenesis, cell signalling and the regulation of gene expression (CN pp. 61–64, 74–80, 82–93).

A protein is formed by the folding of the linear chain of amino acids synthesised from its gene. In theory, this folding produces one single three-dimensional structure for each protein. According to the principle of stereospecificity, a protein would therefore have a stable, ordered, three-dimensional structure which would strictly determine how it functions and the possibilities it would have of combining with other proteins (Wu, 1931; Mirsky and Pauling,

[17] Etymologically stereospecificity means 'solid specificity', i.e. specificity in the material relationships between molecules. The concept of specificity itself comes from Aristotle's philosophy (see chapter 2, §2.2.3).

1936). The first proteins that were studied between 1950 and 1960 indeed complied with this principle (Kendrew *et al.*, 1958). From these individual observations it was extrapolated that, in general, biological molecules recognise each other owing to their shape and their electrical charges, as do locks and keys. Each molecule possesses specific non-covalent linkage sites corresponding to the most stable interactions with other molecules. A group of molecules brought together would spontaneously form the most stable, ordered structure possible, i.e. the one that maximises the number of specific linkages (Fig. 6). Such a process, called self-assembly, was put forward in the first instance to explain the genesis of viral structures (Caspar and Klug, 1962), and was then generalised to all cell structures (Bouck and Brown, 1976; Inoué, 1982).

As Jacques Monod said, in a process of stereospecific self-assembly, "*As in a crystal, the structure of the assembled molecules itself constitutes the source of 'information' for the construction of the whole. These epigenetic processes therefore consist essentially in this: the overall scheme of a complex multimolecular edifice is contained* 'in posse' *in the structure of its constituent parts, but only comes into actual existence through their assembly*" (CN pp. 86–87).

In as far as the three-dimensional structure of proteins depends on their linear sequence in amino acids, which depends in turn on the nucleotide sequence of the DNA, according to this theory, ontogenesis really is the transformation of the genetic information into a material process, complying with the principle of order from order.

However, there is a major point to note. For the spontaneous self-assembly of a structure to be reproducible, the possible interactions between molecules have to be unequivocal or very limited in number, so as to avoid generating too many possible combinations which would prevent the genesis of a unique ordered structure (Kupiec, 1999). Caspar and Klug's models of viral structures (1962) help to illustrate this problem (Fig. 6B). A viral particle is produced by the self-assembly of identical proteins, complying with a repeated basic motif. Each molecule, diagrammatically represented by an irregular polygon, can only combine with another molecule via identical binding sites, symbolised by the homologous sides of

A

B

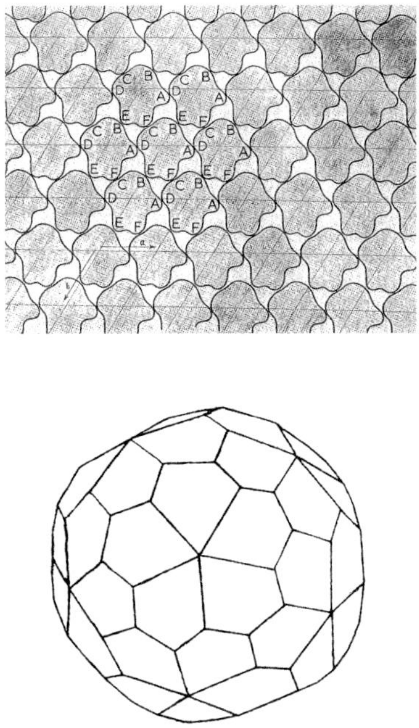

FIGURE 6. The principle of self-assembly. In order to create unique structures, proteins must interact unequivocally. **A**: Example of a single protein that self-assembles to form a viral or cell structure. AD, BE and CF are the only contacts possible. **B**: Example of an icosahedral viral structure. The protein is symbolised by an irregular polygon. No polygon can interact other than through an identical side of another polygon. If these rules were not respected, ontogenesis would not be possible because it would end in multiple structures formed at the whim of molecular encounters. We are grateful to Donald Caspar, Aaron Klug and CSH Press for permission to reproduce this figure (Caspar and Klug, 1962).

the polygons. If the molecules could also be bound by their other sites, a combination of possible interactions would be created and the spontaneous self-assembly of these molecules would no longer produce a unique viral structure. Several forms would be possible, each with a probability of being produced which would depend on the combination possibilities and the frequency of interactions between the molecules.

If the molecules did not follow the unequivocal rules of association, this problem would be even greater in the case of more complex cell structures involving a much larger number of molecules. The possibilities for combination would then be enormous and the probability of forming a functional structure from among all those which are possible would be very slight. Ontogenesis would no longer be a deterministic but a probabilistic mechanism, which would obviously be contrary to genetic determinism.

The idea of stereospecificity therefore involves the unequivocal character of molecular interactions. To form a complex structure, biological molecules have to fit together excluding chance, like the pieces of a puzzle. At this price, we can understand the mechanism of genetic determinism. An organism is constructed gradually from the stereospecific order of the molecules per organisation level, depending on a series of causal determinations running from the genome to the phenotype: 1) the proteins are synthesised from the genes; 2) the proteins assemble stereospecifically to form cellular organelles and cells; 3) due to the exchange of specific signals carried by the proteins, the cells recognise each other and form tissues and organs; 4) the process ends with the organism which was coded in the genome being produced (see Fig. 11; CN pp. 93–94). Each of these levels, called an 'integron' by François Jacob, is produced by the integration of specific interactions of the lower level (Jacob, 1970). Owing to the 'central dogma of molecular biology', the process occurs only in one direction, always from the genotype to the phenotype. It is the genome therefore, that, according to this consensus vision established in the 1960s, holds the power of organisation. The genome is the cause of the phenotype in the strongest sense of the term, carrying the coded representation of the organism, and determines the mechanisms of cell morphogenesis.

This theory helps in understanding the reasoning behind molecular biology research efforts, which consist of systematically isolating genes and proteins and subsequently, sequencing the human genome. Indeed, if all biological phenomena were supported by stereospecific interactions, it should be possible to characterise a protein involved in one of these phenomena and isolate other proteins that interact with it; it should be possible to analyse, therefore, the cascade of molecular interactions that determine it[18] and thus be able to fully explain it. From a protein, it should be possible to reconstitute the entire causal chain underlying the phenomenon. This is why the analysis of a biological process always starts by isolating a protein (or the corresponding gene) implicated in that process. This theory justifies the genome sequencing programme which should lead to our acquiring every bit of information relating to these stereospecific proteins and the organisms they produce.

3.3 Genetic programming and signalling

Stereospecificity is also a key concept for understanding how genetic programming functions. Although this raises a great many problems, the genetic programme was conceived by analogy with a computer program (Longo and Tendero, 2007). According to genetic theory, cells differentiate because the different sets of genes which are active in the various types of cells confer particular morphological and functional properties on them (Morgan, 1934). To explain this, it has been assumed that some genetic information controls how the genome itself functions. There are said to be two types of genes: structural genes encoding for the proteins directly involved in the construction of cells or how they function, such as membrane proteins or enzymes, and regulator genes, controlling the activity of these structural genes. According to the theory of genetic programming, the regulator genes are responsible for the differential expression of genes, because they code for the proteins which act as signals to activate or inhibit the structural genes. Their own regulation during cell differentiation corresponds to the genetic programme of the organism. However, for this theory to be acceptable, it is necessary to explain in concrete terms how these signals, which

[18] This is possible since techniques for finding the molecular partners of a protein are available now.

regulate gene activity, act. The work of Monod and Jacob on the regulation of genes involved in lactose metabolism provided an answer (Jacob and Monod, 1961).

When *Escherichia coli* bacteria grow in a medium without lactose, the enzymes that allow it to be metabolised are not synthesised. If lactose is added to the culture medium, their synthesis increases greatly in less than two minutes. This regulation permitting the bacterium to adapt very effectively to its environment occurs directly at the level of gene activity.

Three structural genes code for the enzymes of lactose metabolism. They are grouped together on a single portion of DNA called the lactose operon (Fig. 7). A regulator gene *i* codes for a repressor protein *R* which represses the activity of the three genes on the operon. Regulation of this operon involves two other portions of DNA, the operator *o* and the promoter *p*, situated upstream of the lactose genes. *R* binds stereospecifically to the operator *o* or to lactose, but these bonds are mutually exclusive. In the absence of lactose (Fig. 7A), *R* fixes on *o* and the enzyme, RNA polymerase, which transcribes the lactose genes, is prevented from passing beyond the promoter *p*. In the presence of lactose (Fig. 7B), *R* binds with it liberating the site *o*. Transcription of the genes can then resume.

Other regulatory models have been described for *Escherichia coli*. In some cases, the regulator protein is an activator that stimulates gene activity. In all these models, however, very precise regulation which excludes chance is thought to be due to the property of stereospecific recognition between the molecules. Because regulator signals act according to an "all or nothing" rule, the genes are either active or repressed.

This regulation logic has been generalised since to the overall way in which the genome functions in multicellular organisms (Monod and Jacob, 1961). The differential activity of the genes in the cell lineages of an organism has been explained by cascade regulation similar to that of the lactose operon. Indeed, in this model, if one of the genes of an operon codes for a regulator protein capable of controlling the activity of other operons in the same genome,

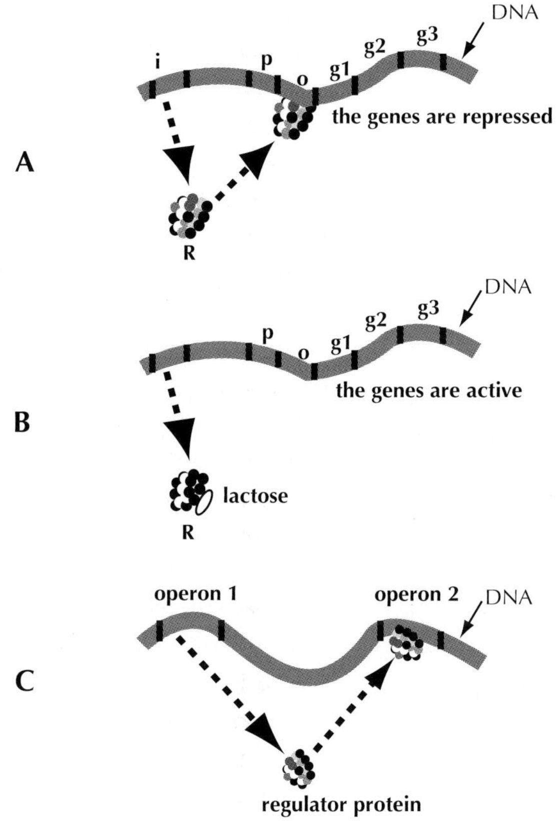

FIGURE 7. The regulation of genes and the genetic programme. A: In the absence of lactose, the repressor fixes on the *o* sequence and blocks transcription of the genes g_1, g_2 and g_3. **B**: Lactose binding to the repressor removes this inhibition. **C**: If a gene of an operon codes for a regulator protein, this induces cascade regulation of several operons (genetic programme).

there is a relationship between the genes which ensures that their activity is coordinated (Fig. 7C). Consequently, programmes relating to the overall activity of the genome can be explained by progressive complexification of the model, involving cascade regulation of all the genes. How they function is based on the property of stereospecificity owing to which regulator proteins act like cybernetic commands activating or repressing genes. In some cases, these regulator proteins could also leave the cell where they are synthesised to enter other cells and exert their effect there. Coordinated differentiation between several cell lines due to this intercellular signalling could thus occur, explaining the global activity of all the cells of an organism by networks of molecular interactions.

However, as with the case of morphogenesis by self-assembly, for signalling and regulation of genetic expression to be effective it is necessary for the underlying molecular interactions to obey the rule of stereospecificity and for them not to generate any possibility of multiple combinations. If that were the case, several possible responses would correspond to a given signal i.e. cell mechanisms would no longer be deterministic. Molecular interactions must therefore be as precise as cybernetic commands directing the living being. As Monod says, it is "*...the huge network of cybernetic interconnections which makes each organism an autonomous functional unit...*" (CN p. 79).

For molecular biology, the organism is therefore still a deterministic machine, though Descartes' old clock has been replaced by a computer.

4

The Contradiction in Genetic Determinism

SUMMARY. In contrast to theoretical predictions of genetic determinism, experimental data obtained over the last forty years show that interactions between biological molecules are not specific and are immensely varied, with one molecule being able to interact with a large number of partner molecules. There are several causes for this phenomenon. The same amino acid sequences corresponding to the binding sites between molecules are present in a great many different proteins, which produces a very large number of possible combinations of potential interactions, increased even further by less powerful interactions involving sequences said to be 'degenerate'. However, there is a more radical cause of non-specific interactions. It is now known that many proteins do not have an ordered three-dimensional structure necessary for producing specific interactions. They can adopt a multiplicity of conformations permitting a huge number of interactions, which generate different macromolecular structures or cascades of interactions. This directly challenges the principle of order from order of genetic determinism, according to which biological organisation is supposed to arise from an underlying molecular order. All the molecular cascades are interconnected one with another, and simply mapping the networks they constitute cannot explain the way a cell functions. Molecular networks must themselves be subject to some kind of regulation which remains to be elucidated. To solve this problem, biologists have inferred that the overall structure of the cell restricts the combination of molecules, but such an explanation is holistic. It comes down to asserting that, in contradiction with molecular biology and genetic determinism, biological organisation does not arise from the properties of molecules, but from cells as a whole.

A large amount of work has been done since the beginning of molecular biology research to isolate stereospecific proteins. A great many molecules have been characterised, such as the transcription factors involved in regulating gene expression and the signals permitting the transfer of information. These discoveries have resulted in major progress, yet at the same time, a problem has arisen. When the interactions of these proteins were analysed, instead of finding single or a well-defined limited number of interactions, it came to light that they do not exhibit the character of specificity expected, but on the contrary, are conducive to a great variety of interactions. Many proteins can interact with a large number, several tens, or even hundreds, of partner molecules, generating an immense number of possible combinations. This is confirmed for the proteins implicated in all biological phenomena and there are countless cases of molecular non-specificity described in the literature. It is not our aim here to carry out an exhaustive review, which would be impossible, but to illustrate it using a few examples from the major areas of cell physiology, concentrating on the signalling and regulation of gene expression. We will also analyse the causes of this non-specificity and its consequences for a theory of biological organisation.

4.1 The non-specificity of biological molecules

4.1.1 *Non-specificity in metabolism*

The precision of biological processes relies for the most part on the precision of enzyme reactions, yet there are many examples of non-specificity in this area. Non-mutated enzymes can act on many substrates, which may be exogenous substrates, as in the case of human carboxylesterase 1. This enzyme is known for metabolising heroin and cocaine but it acts on numerous other poisons such as sarin, soman and tabun as well (Bencharit *et al.*, 2003).

The multiple substrates of an enzyme are often also endogenous substrates produced by cell metabolism in prokaryotes and eukaryotes. This subject has already undergone analytical review (D'Ari and Casadesus, 1998). We shall only mention a few of the elements here as examples. In some cases the different activities of an enzyme are useful to the cell, for example isoleucine and valine are synthesised by the same enzymes, and, similarly, four transaminases which have crossed specificities catalyse the formation of seven amino acids. But in other cases multiple reactions do not seem to be useful to the cell. One example is the oxygenase activity of the enzyme rubisco, which wastes oxygen during oxidative hydrolysis of ribulose diphosphate.

4.1.2 *Non-specificity in the immune reaction*

The antigen-antibody reaction was considered as the perfect model of specific interaction, permitting the immune system to effectively resist infection or contamination. But this dogma has now been demolished. The antibodies produced against a particular antigen exhibit cross-reactions with other antigens. This non-specificity is due to the flexibility of the antibody binding domains (Manivel *et al.*, 2000; Mundorff *et al.*, 2000; Garcia *et al.*, 1998). Cross-reactions have been observed, among other things, between ovalbumin, bovine gamma globulin and bovine serum albumin antibodies (Sperling *et al.*, 1983) and between butyrophilin and myelin antibodies (Guggenmos *et al.*, 2004). In the same way, antibodies against either (4-hydroxy-3-nitrophenyl)acetyl, p-azophenylarsonate or a synthetic peptide react equally with a multitude of ligands present in banks of random peptides[19] (Manivel *et al.*, 2002).

[19] These banks are obtained by randomly synthesizing a peptide of a given size. Each amino acid can be one of the 20 possible. This technique generates an enormous population of different molecules.

Non-specificity in cellular immunity has also been demonstrated on countless occasions. The receptor carried by a T-lymphocyte is capable of recognising antigens different from the one that induced the immune reaction (Amrani *et al.*, 2001; Hausmann *et al.*, 1999; Dutoit *et al.*, 2002).

4.1.3 *Non-specificity in cell signalling*

Cells receive various signals from their environment. In bacteria, chemotactic signals indicate a source of food or danger, while in multicellular organisms, signals encourage the multiplication or differentiation of cells. In these signalling processes, the first step generally involves the binding of the signal carried by an extracellular chemical ligand with the extracellular domain of a receptor molecule located in the cell membrane. This binding activates the intracellular domain of the receptor which then triggers a cascade of molecular interactions inside the cell, transducing the signal (Fig. 8). The crossed reactions between antigens and T-lymphocyte receptors that we mentioned in the previous section are not exceptional. Although the cells have to respond precisely to the signals they received, non-specificity affects the receptor binding to its extracellular ligand just as much as it affects the reactions that transduce the signal within the cell.

For example, the bacterium *Escherichia coli* uses only four receptors to respond to pH, temperature, and about 50 chemical substances (Bray, 2003; Ames *et al.*, 2002). In the case of the mammalian epithelial growth factor receptor, at least six different ligands have been identified (Schweitzer and Shilo, 1997; Carpenter, 2000). In the same way, chemokines are involved in thymocyte migration and other cellular functions, particularly the production of blood cells, with more than 50 chemokines having been identified for only 16 receptors. Each of the 50 chemokines can interact with one or more of these 16 receptors and, conversely, each receptor can interact with several chemokines (Broxmeyer and Kim, 1999; Fu and Chen, 2004).

The situation is similar as regards intracellular reactions involving receptors and the interaction cascades they induce in the cells.

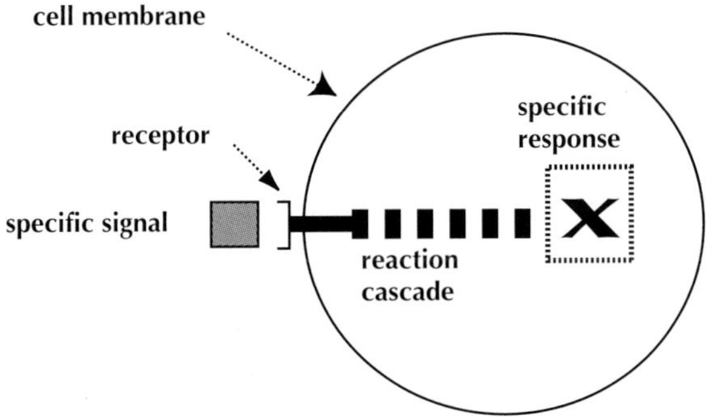

FIGURE 8. **Transduction of the signal.** A specific signal binds to the extracellular domain of its receptor situated in the cell membrane. This binding activates the intracellular domain of the receptor which in turn activates a cascade of reactions within the cell, terminating in a specific response X, for example, activation of one or more genes.

They frequently bring into play kinases or phosphatases which recognise their substrate by very short amino acid sequences present in many proteins. This is the case, for example, with tyrosine kinase receptors. Initially, using classic biochemical methods, eight proteins that could bind with two different receptors were described (Kazlauskas, 1994), but these data underestimated the situation. Although they have not been precisely counted, a very large number of potential partner molecules has since been discovered (Hunter, 2000; Castagnoli *et al.*, 2004).

Transduction of the signal often ends with nuclear proteins which activate or repress certain genes, and countless molecular interactions occur here, too. The nuclear receptors of oestrogen hormones respond to signals from xenoestrogens or from growth factors. In addition they interact with at least 25 proteins involved in a variety of cell functions (Moggs and Orphanides, 2001).

4.1.4 *Non-specificity in the control of gene expression*

The interactions between chromatin proteins which control gene expression and their binding sequences in DNA are also non-specific.

These sequences are only six to twenty nucleotides long, and numerous copies of them are present in the genome, such that many interactions may occur.

Hox genes are an illustration of this situation as they determine several stages of embryonic development. The transcription factors they encode activate numerous genes implicated in differentiation in the early embryo stage or in the limbs of vertebrates and insects. These proteins do not however show any specificity in their binding to DNA. The sequences they recognise are only six nucleotides long and are therefore very frequent in the genome (Gehring *et al.*, 1994). As a result of this, they are capable of binding with any gene[20] *in vitro* whereas they only bind with a limited number of genes *in vivo* (Carr and Biggin, 1999; Biggin, 2001).

In weaker interactions transcription factors also recognise what are called 'degenerate' sequences, which only differ from the sequences of maximum affinity by one or more nucleotides. These degenerate sequences are also repeated many times in the genomes of multicellular organisms, thus increasing the possibilities of interaction (Zhang *et al.*, 2006; Bendall *et al.*, 1993).

There are even more spectacular examples. MeCp2 represses the activity of the genes recognising the methylated CG dinucleotide. This target is present 40 million times in a mammalian genome whereas there are only a million MeCp2 molecules (Nan *et al.*, 1997).

Thus, as with protein-protein interactions for signal transduction, there is a huge number of potential protein-DNA interactions.

4.1.5 *Overall non-specificity of protein networks*

The previous examples of non-specificity were obtained from studying individual proteins. If each stage in every domain of cell physiology is subject to similar multiplicity, the total number of combinations of possible interactions for a whole cell must be enormous. This has now been verified experimentally. Networks of interactions between proteins have indeed been studied globally in

[20] Or with the regulator regions of these genes.

several organisms such as yeast, *Drosophila* and humans and all the interactions which may occur in a cell have thus been mapped (Bork *et al.*, 2004). These large scale proteome studies are not yet absolutely complete but the results they have already provided are significant. Protein interaction networks are constructed around a central hub where connection density is the strongest. This area is composed of proteins which can bind to approximately a hundred other partners and constitute about 10% of the network. Their number is therefore in the order of 10^3. The connectivity of other proteins located on the periphery of the network is much lower, but, overall, the average connectivity of proteins over the whole of the network is between five and ten. These data suggest that all protein interaction pathways implicated in metabolism, signalling or gene transcription are potentially interconnected, with a very great number of contact points (Barabasi and Oltvai, 2004; Albert, 2005).

These wide-ranging studies confirm the results obtained from those restricted to particular proteins, and demonstrate that an immense number of combinations of potential molecular interactions must exist in cells. Indeed, the play of multiple interactions very rapidly provokes a 'combinatorial explosion'. In his book *The Music of Life,* Denis Noble estimated the number of potential interactions between the 25000 genes of a mammalian genome and reached similar conclusions (Noble, 2006). To give an idea, let us do a very simple calculation and consider a cascade of 20 sequential protein interactions. The first protein can interact with one of seven other proteins (first interaction), each of these seven potential proteins can in turn interact with one of seven others (second interaction), each of the 49 proteins of the second interaction can in turn interact with one of seven other proteins (third interaction), and so on up to the 20th interaction. The number of possible combinations for producing this cascade of 20 interactions is in the order of 10^{17}. In comparison, one must remember that in a mammal the number of cell types is in the order of 10^2 and the total number of cells, around 10^{12}. One should also keep in mind here that this is only a rough calculation leading to an underestimate. Obviously there are more than 20 interactions in a cell!

4.2 The causes of molecular non-specificity

The principle of stereospecificity, on which genetic determinism is based, which implies that relationships between biological molecules are unequivocal, or very limited in number, does not therefore comply with experimental evidence. Biological molecules are capable of multiple interactions and the number of combinations for them in one cell is enormous. We have described the non-specificity of proteins. About ten years ago, the role of 'interfering' RNAs in regulatory processes was revealed, but in Richard Burian's analysis their action is subject to the same problem as experienced by proteins. Their interactions are not specific and generate a huge number of combination possibilities (Burian, 2008).

The ontogenesis of a single living structure cannot therefore be the result of a process of self-assembly bringing only molecular affinities into play, because several structures are possible due to this number of combinations. Other factors evidently have to come into effect so that the number of combinations is limited. Before analysing this problem in the face of genetic determinism, and envisaging its theoretical consequences, we must discuss the causes of molecular non-specificity. We will see that they directly contradict the principle of order from order, i.e. the idea that there is a biological order intrinsic to the living organism which is carried by proteins.

4.2.1 *The multiplicity of interaction domains*

Proteins interact via interaction domains, which are structural motifs, corresponding in general to sequences that are 40 to 150 amino acids long (Hunter, 2000). There are a great number of these domains corresponding to the different amino acid sequences. One cause of non-specificity comes from the fact that the same domain may be carried by many proteins.[21] The sequence coding for the

---

[21] Even if the context in which a domain is inserted partly restricts its possibilities for binding with other proteins.

domain called SH2 allowing phosphotyrosine binding is present 115 times in the human genome, while the sequence of domain SH3 of the tyrosine kinase CsK is present 253 times (Pawson and Nash, 2003). In addition, these repeated domains often recognise very short binding sequences which are only four to ten amino acids long. For this reason these sequences are themselves present in a large number of proteins, which are just as many potential molecular partners. The domain SH3 thus recognises the amino acid sequence P-X-X-P.[22] Countless other interaction domains favourable to such combinations have been identified (Castagnoli *et al.*, 2004).

4.2.2 *The plasticity of interaction sites*

Another reason for molecular non-specificity puts paid to the idea we have of there being one molecular interaction between two well-defined entities. Not only are the same interaction domains present in many proteins, but a single protein domain can bind to different ligands. The domain of SMAD proteins called MH2 provides an example. These proteins are used in transducing signals between the cell membrane and the nucleus, where they modulate the activity of several genes. During this transfer their MH2 domain interacts with many partners carrying different binding sequences (Pawson and Nash, 2003). This phenomenon causes the number of combinations of possible interactions to be multiplied and challenges the static view of stereospecificity. Indeed, for a single domain the possible ligands can be very different, in form, size and amino acid composition. The number of arguments indicating that this phenomenon is due to a protein interaction site not being a static entity, but a dynamic one, is growing. Its three-dimensional structure is not rigid but flexible. It constantly changes its configuration. A protein in solution would in reality be a population composed of a mixture of several conformations in dynamic equilibrium, each with a particular potential 'specificity'. Structures deduced by crystallisation are in fact only frozen images which eliminate this diversity

---

[22] P = proline and X = any amino acid.

of conformation. Seen from this perspective, it is not the pre-existing structure of the protein which determines its future interactions but the ligand, which stabilises one of these conformations and alters the equilibrium of the population (Ma *et al.*, 2002).

4.2.3 *Molecular disorder*

There is an even more radical cause of molecular non-specificity. We have already emphasised the fact that molecular biology is based on the idea that proteins have a well-defined three-dimensional structure, and that macroscopic biological organisation arises from this microscopic order. This dogma has now been demolished. It has been shown that a large fraction of proteomes correspond to proteins which contain intrinsically disordered regions, incapable of generating secondary and therefore stable, three-dimensional structures by themselves. The disordered regions comprise in general more than half of each of these proteins and often their entirety. They are not of secondary importance. On the contrary, proteins only acquire a functional structure when the disordered regions are stabilized by interaction with another molecule. Owing to their great plasticity, they can interact with a large number of partners adopting a different configuration and function in each case (Wright and Dyson, 1999; Dunker and Obradovic, 2001; Dyson and Wright, 2005; Dunker *et al.*, 2005). For example, HMGA is a nuclear protein which is intrinsically totally disordered. It has an important role in structuring chromosomes and chromatin, and in the transcription of at least 45 genes. To perform this role it interacts with the chromosomal structures, the nucleosomes, and with at least 18 different transcription factors. In each case interaction with a different partner confers on it a particular functional structure. Another well-known example involving a fundamental biological function is protein p21 which is known for its essential role in the cell cycle. It inhibits a variety of molecular (cyclin-Cdk) complexes thanks to variable conformations stabilised by the interactions. These are not isolated cases. Today, we know of hundreds of proteins that can modify their structure and function through such a structural interaction mechanism. In very many instances, these interactions seem to be determined only by the probability of encounters between partner proteins (Beckett, 2004). Their amino acid composition, hydrophobic nature and electrical charge give disordered proteins a characteristic signature which really differentiates them from structured proteins. It is possible, with appropriate algorithms, to analyse entire genomes or banks of protein sequences and determine the proportion representing disordered proteins, and we have thus been able to measure the overall involvement of disordered proteins in various cell functions. They make up 36% to 63% of genomes in eukaryotes but only 7% to 33% in prokaryotes and archaebacteria. Protein disorder is therefore positively correlated with multicellularity (Dunker *et al.*, 2000). It is also significantly increased in signalling proteins and those implicated in cancer (Iakoucheva *et al.*, 2002), in transcription factors (Liu *et al.*, 2006), and in the "hub proteins" of protein networks (Haynes *et al.*, 2006). These studies demonstrate that protein disorder is not a marginal phenomenon: it is surprisingly present even in cell signalling and gene transcription.

We have to acknowledge, therefore, that the existence of these proteins radically challenges the conventional idea we have of the relationship between a gene and the structure and function of a protein. Their structure does not depend in a deterministic way on their sequence coded in the DNA, but on their encounters within the cell. Their structure and function are not therefore written, pre-existing and unalterable, in the genome, but are produced by cellular processes in real time. Now, it is not possible to envisage genetic programming as precisely determining intermolecular encounters. Certain data even strongly suggest that a certain randomness inevitably comes into play here. In an extreme case, the same intermolecular encounter can produce different effects because the two partner molecules may interact in a variety of ways inducing different conformations and functions. The choice between these options seems then, to be probabilistic (Haarmann *et al.*, 2003).

4.2.4 *Specificity is not an experimental concept*

Finally, there is an epistemological problem related to using stereospecificity as a concept. Proteins cannot be specific quite simply because the concept is not relevant for describing experimental reality (Kupiec, 1999). It automatically imposes an arbitrary order on the way we look at the natural world, even if this order does not actually exist. It is a qualitative notion, in fact, whereas in practice we analyse molecular interactions with quantitative parameters. Specificity follows the 'all or nothing' rule and according to the way of thinking it imposes, two molecules either are or are not specific to each other. Reality does not however comply with this Aristotelian logic and its ordered way of dividing up the world in a discontinuous manner. A molecular interaction is measured by the equilibrium constants for the complex that the molecules form, no interaction being absolutely stable. What is measured is the longer or shorter average life span of the complex between two dissociation events. The greater the affinity, the more stable the complex will be and the longer its average life. A given molecule can always interact with many partners, with stronger or weaker variable affinities. The experimenter is obliged, owing to this continuous, quantitative character of molecular affinities, to set a threshold below which he will consider the interaction as non-specific, but that does not mean that weak interactions do not exist or that they do not occur in the organism.

This approach is subjective, and leads to a bias in our appreciation of reality and to a contradiction. Nothing gives us leave to declare *a priori* that a weak interaction has no biological effect. It may even be that a weak interaction repeated often would have more biological effects than a strong interaction that occurs rarely. Even if weak interactions do not have direct physiological consequences, the simple fact that they occur means that they come into competition with strong interactions and affect their kinetics. They therefore also contribute to determining the state of a biological system. Despite this, we always operate arbitrary selection, which leaves weak interactions to one side.

It is possible, to be sure that an interaction is really relevant, to confirm that it occurs *in vivo* in its original cellular context, so as to leave aside interactions which are only detected *in vitro* (Krause *et al.*, 2004). This strategy is also biased. We are no longer measuring the intrinsic capacity of the protein to form bonds due to its physical structure but rather the bonds that occur in a particular context. Other factors present in the cell, such as molecular cofactors or the structure of the cell, always influence the spectrum of bonds detected *in vivo*, by promoting some interactions and forbidding others.

The concept of specificity therefore leads to underestimating the physical interaction possibilities of biological molecules because it does not encompass the quantitative and continuous aspects of this phenomenon.

4.3 The consequence of molecular non-specificity: Return to holism

4.3.1 *The network won't work*

We thought we could explain mechanisms of regulation by linear cascades of clearly defined molecular interactions, but molecular non-specificity makes them a lot more difficult to understand. We come up against the fact that interaction cascades are interconnected one with another. Two specific examples illustrate this problem.

The first shows how a signal can activate several different cascades which diverge. The Ras protein plays a major role in controlling cell multiplication and also influences other processes such as differentiation and apoptosis. It acts as a relay in the transfer of various extracellular signals such as growth factors, cytokines and hormones. A linear interaction cascade was first characterised which, from the cell membrane to the nucleus, successively involved the protein Raf and a series of kinases, ending in activating the transcription factor, Elk-1. The causal chain explaining the role of Ras in cell multiplication was believed to have been elucidated.

This simple succession of events became complicated, however, when it was discovered that Ras did not solely interact with Raf but with at least eight other effectors involved in several cascades activating many transcription factors. Because of these multiple activations, the Ras protein has pleiotropic effects, and its action on cell multiplication is a much more complicated process which must depend on precise equilibrium between all these effects (Campbell *et al.*, 1998). It has to be acknowledged, therefore, that the initial explanation is no longer adequate, and a new, very difficult question arises from this example: if cell functions depend on equilibrium between the activity of several signalling pathways, how is this equilibrium controlled differentially and specifically in the different types of cell in order for them not to perform the same functions? We have to depart, in fact, from the initial theoretical context based on specificity. The necessity for this can be demonstrated with a little simple calculation. Let us take the one we did in section 4.1.5 of this chapter, but instead of considering a cascade of 20 interactions produced in an overall cell network, let us simply take a cascade of four interactions, which would describe a signalling pathway from the initial protein signal to its target. This is a sensible size of cascade for doing our calculation although in certain cases real cascades may be longer. At each step in the cascade, there are again seven interaction possibilities with different proteins for each protein. In this case, the combination possibilities are such that the signal may activate (or repress) 2401 different targets. If there is one more step in the cascade, this number rises to 16 807. If the cascade involves six interactions, which is still a reasonable number in view of the size of actual cascades, we are faced with 117 649 potential interactions. How, from the point of view of deterministic functioning based on stereospecificity, is the signal going to be directed to its specific target among these thousands of potential interactions? One answer is to say that there is a set of molecular targets for each signal and that this set is what is specific to the signal. Yet this answer does not hold water either. Let us go back to our calculation. In a mammal there are about 250 different types of cells. According to the theory of genetic programming, at least one signal corresponding to the induction of each of these cell types is required. However, cells must also be subject to a vast number of signals corresponding to other cell functions such as multiplication and apoptosis, or physiological functions controlled by hormones. We might therefore estimate that the minimum number of specific signals necessary in a mammal is 1000, and there again, this is still a considerable underestimate. Even if this were the case however, if these 1000 signals each activate a cascade of four interactions, that means that they must activate a total of $1000 \times 2401 = 2.4 \times 10^6$ specific targets. Now we know that in one cell there are about 10^4 genes corresponding to roughly the same number of proteins. In terms of specific regulation, we once again come up against impossibility. Far too many specific targets are necessary in relation to the possibilities offered by one cell. Some signalling pathways are bound to be used by several signals. The second example of interconnected interaction cascades shows that this is indeed the case.

There are indeed relatively few signalling pathways in a cell, compared with the enormous number of signals that the cell can receive and situations with which it is confronted. Because of the multiplicity of molecular interactions, a single cascade of signals can produce different effects. The same pathways are used by different signals for transporting their information and achieving responses appropriate for the cell. The yeast *Sacharomyces cerevisae* uses three kinases, Fus3, Hog1 and Kss1 to respond to the sex pheromone, to osmotic pressure and to induce filament growth. The three pathways which activate these kinases share several portions made of the same proteins and yet, depending on which signal it is that activates them, only one of the three responses is produced (Schwartz and Madhani, 2004).

However, this once again raises the question of the specificity of the signal. The problem can be simply represented in a diagram. Three signals A, B, C converge to use the same signalling pathway in a non-specific manner, then diverge and induce three specific responses A′, B′, C′, respectively (Fig. 9). Why does each signal induce a unique response instead of the three responses possible?

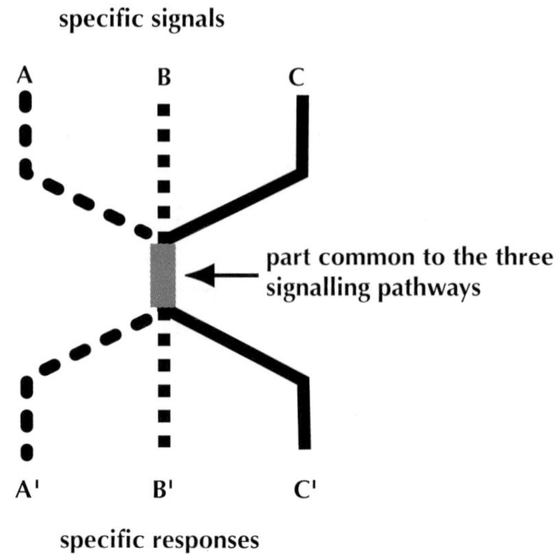

Figure 9. The problem of the specificity of the signal. Three signals A, B and C activate three responses in a cell, A′, B′ and C′ respectively. Yet they use in part a single signalling pathway. How is the specificity of these signals maintained?

We could easily generalise these examples and show that the same question arises for the regulation of gene expression. How can genetic programming work if the interactions between the regulator proteins and their target sequences in the DNA are not specific?

Systematic study of proteomes has shown that all the signalling pathways of a cell are interconnected (see this chapter, §4.1.4). When the molecular complexes of a cell are isolated and analysed, at least 37% of the proteins are found in several complexes performing different functions (Krause *et al.*, 2004). This is therefore a general problem in the way cell networks function. How can a particular signal induce a specific response instead of activating all the functions of the cell and causing all the effects possible to be scrambled? How does the cell function in these conditions? It is usually suggested that the functioning of molecular networks is itself subject to spatial and temporal dynamics, and therefore the same parts of a network would not be activated at the same time at a different point in the cell, so that specific responses are thus generated (Kitano, 2002; Levchenko, 2003; Prill *et al.*, 2005). Before describing the mechanisms alleged to control the activity of the networks, the following must be stated. Under the principle propounded by genetic determinism of order from order, the regulation and macroscopic organisation of biological systems is supposed to be explained by the structure of the protein networks, itself resulting from molecular stereospecificity and genetic information. This is however not the case. Regulation of the networks themselves must now be put forward to explain their specificity.

4.3.2 *Negating the principle of order from order*

The mechanisms put forward to explain how a molecular network generates specific behaviour despite the non-specificity of the proteins which compose it have been the subject of in-depth descriptions (Dumont *et al.*, 2002; Schwartz and Madhani, 2004; Komarova *et al.*, 2005). We shall look at the essential points here. We shall see that these mechanisms, unanimously accepted by molecular biologists, reintroduce holism, yet this absolutely contradicts the principles of genetic determinism.

The sequestration of proteins consists of limiting contact between proteins, in order to prevent certain interactions from occurring and only let those that occur exert a supposed specific effect. It is itself the result of several mutually non-exclusive mechanisms.

— Spatial compartmentalisation: proteins are not uniformly distributed in a cell. They are preferentially located in certain compartments such as the nucleus, the cytoplasm, the membranes or other organelles. Compartmentalisation therefore prevents interactions between physically separated molecules.

— Temporal separation: some proteins are not present at the same time at the same place in the cell because they are not expressed with the same kinetics. Their interacting is thus avoided.

— Micro-compartmentalisation: there exist proteins called 'scaffold' proteins which bind to the various proteins of a single signalling pathway. The latter are then compelled to react with each other because of their proximity. The signalling pathway is thus preferentially activated.

Signal transduction via a combination of intracellular pathways is another mechanism conferring specificity. Two signals share a mutual pathway but at the same time they each activate other different pathways. In this case, specificity is conferred by the combination of different pathways activated by each signal. In the same way, in gene expression it is the combinations of transcription factors which appear to ensure regulation.

Crossed inhibition may equally restrict the effects of molecular non-specificity, with an element unique to one of the two pathways possibly inhibiting an element unique to the other, even though they have elements in common. From the moment there is an imbalance in favour of one of the two pathways, for example owing to the presence of a scaffold protein, this pathway will totally inhibit the other.

Finally, the intensity of the signal could also play a part. Depending whether a pathway is activated by one signal or another, the magnitude and period of activation of a single intracellular pathway could be different and thus end in producing different effects.

All these mechanisms are supported by solid experimental data which explain how appropriate regulation occurs despite molecular non-specificity.

However, there is no denying that this shakes genetic determinism to the roots and that we have arrived at a contradiction. Indeed, for all the suggested mechanisms to be effective, it has to be assumed that cellular organisation and a state of macroscopic differentiation already exists to ensure compartmentalisation and the very precise expression of certain proteins. It is this differentiated state, specific to one cell, which must explain why certain molecular interactions occur specifically in this cell and not in others. Yet in genetic determinism, the macroscopic state of a cell is precisely what the molecular interactions are supposed to determine (Fig. 10) and what a theory of ontogenesis must explain. We also arrive at the idea that the effect of a signal does not depend on its

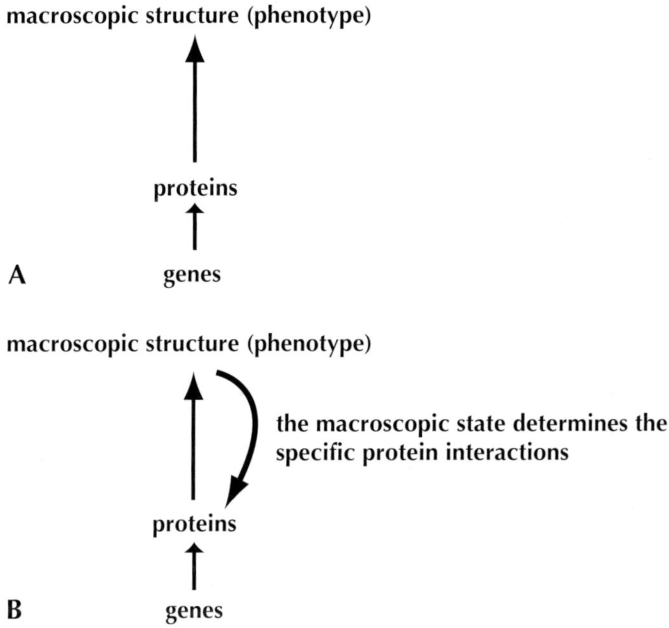

FIGURE 10. **The contradiction in genetic determinism. A:** According to genetic determinism, protein interactions determine the macroscopic state of cells (their phenotype). **B:** It is this macroscopic state which forces the proteins to behave in a specific way. Organization arises therefore from the phenotype (from the global structure) and not from the genes (from the genotype) and proteins. This completely upsets the founding premise of genetics!

intrinsic molecular nature but on the global state of the cell, which permits it to propagate along a specific pathway (Dumont *et al.*, 2002). Although this is very surprising in the context of the reductionist paradigm, this view is furthermore confirmed by a spectacular experiment analysed by Soto and Sonnenschein (2006). Erythropoietin (EPO) and prolactin receptors activate transduction pathways that have several proteins in common. If the normal EPO receptor is replaced in erythroid cells by a prolactin receptor, these cells differentiate into red blood cells when they are stimulated by prolactin (Socolovsky *et al.*, 1998). Prolactin therefore induces the signal normally provided by EPO. This hormone does not normally play any part in red blood cell differentiation, however.

In this experiment, it was the state of the erythroid cell which determined the effect of the hormonal stimulation and not the molecular nature of the hormone.

The study of molecular interactions has therefore completely upset the causal explanation which contradicts the genetic determinism principle of order from order. We are faced with a paradox: the macroscopic characteristics of cells are what determine their organisation and properties at molecular level, not the reverse! Geneticists introduced the genotype/phenotype dichotomy postulating that the genotype determines the phenotype. It has now become necessary to evoke the phenotype to explain the action of the genotype (Kupiec, 2001).

Although all this work was performed by biologists working under the reductionist paradigm, holism, which denies the basic foundations of molecular biology, is back with a vengeance. For the latter to stay theoretically consistent, this contradiction has to be resolved. As the experimental facts indicate, we need a theory integrating the influence of macroscopic structures. Holism has for a long time been pushing the importance of this level of organisation to the fore. In the next chapter we shall analyse whether it can constitute a valid alternative to genetic determinism.

5

Self-organisation Does Not Resolve the Contradiction in Genetic Determinism

Summary. There are numerous variants of holism upheld by philosophers, physicists and biologists. While each has its own special aspects, together they form a real current of thought, with the common characteristic of denying the first principle of science, the latter arising not solely through experimental methodology but also due to the philosophical revolution which abolished animism. The idea of matter animated by a final cause, which was supposed to be inherent to it, was abandoned, for it to be seen as inert and influenced exclusively by external causes. Holism, in contrast, reintroduces animism. It presupposes matter creating organised wholes corresponding to levels of increasing complexity (atoms, molecules, cells, organisms etc.). In this creation, at each level, properties would spontaneously emerge, irreducible to those of lower levels. This model of a world stratified into hierarchical levels, constructed from lower levels, is common to holism and genetic determinism, and the two theories allege that this expresses a real order immanent in the world and living organisms. In the second half of the 20th century, the creator principle of holism took the name of self-organisation and, to account for it, several authors have tried to suggest models applied to physical or biological phenomena. These models do not resolve the contradiction in genetic determinism. They are deterministic models with noise founded on the stereospecificity of the molecules. They themselves contain a contradiction which saps at the foundations of holism. The local properties of the elements of the systems (the cells, the molecules) are not enough to explain their organisation. For this reason, self-organisation models are obliged to include the

> action of external global constraints. Real organisation phenomena are processes of hetero-organisation, not of self-organisation.

This chapter will attempt to analyse holism.[23] First of all, we will recall the founding principles of modern science, then analyse philosophical holism in order to understand in what respect it is opposed to it. Finally, we shall study the theories of self-organisation, and see that they do not provide any solution to the problem posed by the non-specificity of biological molecules.

5.1 The scientific principles

Modern science grew up in the 17th century on the basis of several principles, the most well-known of which is having recourse to experimental method. Knowledge is constructed through dialogue with nature, so every hypothesis should be formulated from observed facts and subjected to experiment. This is the aspect immediately mentioned to differentiate science from earlier scholarly practices. However, it is a simplification that fails to take account of another aspect which is just as important. Although experimental method is essential to scientific practice, it is not enough, as the latter cannot be reduced to methodology alone. Our predecessors in the Middle Ages and Antiquity were not so naïve as to believe that one can assert something without that assertion being logically expressed and conforming to experience. On the contrary indeed, Aristotle created a logic that we continue to use today and pre-scientific discourses constantly resorted to arguments based on observation (Lenoble, 1969). Scientific practice has developed enormously since the 17th century, because a true philosophical revolution occurred which accompanied the development of experimental techniques. Animism and finalism were rejected to make room for a new conception that Jacques Monod called the 'postulate of

[23] The Greek word 'holos' means 'whole', 'all', 'entire'.

the objectivity of nature' (CN pp. 30–31). We stopped believing, at that time, in matter animated by a final cause, with an aim supposedly inherent in it, to conceive of it rather as inert and influenced exclusively by external causes.

Aristotelism, which was the dominant mode of thought before the scientific revolution, assumes that there is a natural order intrinsic to the world. Each thing is said to have a principle of movement or change which forces it to comply with its essence, i.e. to bring about its finality. For a physical object, this means inherent and spontaneous movement towards its natural place of rest. For example, light objects are supposed to rise upwards and heavy ones fall downwards. This is a general system of thought which does not exclusively concern physics, according to which every thing that exists has an essence that determines its behaviour (existence). This system of thought collapsed between the 14th and 17th centuries when the principle of inertia was formulated, abolishing finality and asserting that only external causes act on a body, the latter possessing no activity of its own guiding its fate. This is the principle underlying all modern science.

Indeed, since the existence of things is not determined by intrinsic essence but by the external influences to which they are subject to, there can be no order immanent in the world. It is constructed 'here and now' during all the various kinds of processes which occur there. From this there arises the need to experiment because, in order to understand a phenomenon, one can no longer content oneself with defining the essence of things, as was possible in Scholasticism. It has to be analysed by experiment.

To really grasp this major point, let us take the example of the falling stone. Once the essentialist has seen that it falls, and has asserted that it falls because falling complies with its finality (its essence), there is no longer any mystery to the phenomenon and the explanation is enough in itself, for all time. Each occasion when the initial observation of the stone falling is repeated only serves to confirm this. If the stone does not fall, because it is prevented by an obstacle for example, that does not invalidate the finalist explanation. It just shows that bringing about its finality has been impeded.

This is therefore a closed system of explanation, and in this sense, it is perfect and cannot be faulted. On the other hand, if the stone is inert in itself, what makes it move must be analysed. We are then obliged to hypothesise about the causes (the forces) which act on the stone, and check whether those hypotheses are right using an experimental setup based on a prediction which goes beyond the simple observation of the stone falling. However, as the experimental setup can always be improved by new predictions and technical developments, scientific explanation is never final. It can always undergo new tests and be faulted, necessitating new hypotheses and new experiments. Unlike the essentialist explanation therefore, it is imperfect and open to its own transformation. It is this imperfection that enables it to progress.

In the 20th century, theories of physics profoundly changed. The theories of relativity and quantum physics broke with the deterministic mechanism of the 17th century, but these developments did not mean abandoning the principle of the objectivity of nature and returning to animism. Physics did not discover a new hidden order immanent in the world. On the contrary, as Schrödinger so well explained it, order, for statistical physics, is a subjective approximation (see chapter 3), while for quantum physics, it is a probabilistic theory which has made a fundamental principle of indeterminism. For contemporary science, what is snugly concealed in the depths of nature is not a new finality, a hidden order or some kind of determination, but randomness and indetermination. The anti-essentialism of classic science has been made even more radical, from this point of view.

In biology, the principle of the objectivity of nature led Bernard to elaborate the concept of an internal environment. Indeed, as he explained in *An Introduction to the Study of Experimental Medicine* (1865), its prime aim is not, as is often believed, to define the individual in his autonomy in relation to the external environment, but to allow the development of experimental physiology founded on principles analogous to those of physics and chemistry. The internal environment of a living organism consists of all the conditions which act within the organism on its parts (organs, cells, molecules), causing them to react to the stimuli they receive.

> "*In a word, vital phenomena are the result of contact between the organic units of the body with the inner physiological environment; this is the pivot of all experimental medicine. Physiologists and physicians gain mastery over the phenomena of life by learning which conditions, in this inner environment, are normal and which abnormal, for the appearance of vital activity in the organic units; for apart from complexity of conditions, phenomena exhibiting life, like physico-chemical phenomena, result from contact between an active body and the environment in which it acts*" (ISEM p.76).

Through the concept of an internal environment, we can understand that the phenomena of living organisms are analogous to those of physics and chemistry, that living beings are active, even though their parts are inert in themselves, like non-living matter. It thus removes the need to resort to finalism or vitalism.

> "*In any organic environment, the substances created by animals and vegetables are much more changeable and less stable, but still they are inert and exhibit their properties only as they are influenced by agents outside themselves* (… …) *Therefore, as has already been said, we must not set up an antagonism between vital phenomena and physico-chemical phenomena, but on the contrary, we must note the complete and necessary parallelism between the two classes of phenomena. To sum up, living matter is no more able than inorganic matter to get into activity or movement by itself*" (ISEM pp. 78–79).

We shall come back, in chapter 6, to the very important consequences of the concept of an internal environment not only for experimental method in physiology, but also for constructing a theory of biological organisation (§6.1.2, 6.1.3). For the moment, we shall content ourselves with noting that the development of molecular biology later in the 20th century obviously does not challenge the principle of the objectivity of nature. On the contrary, studying living organisms by physical and chemical methods and concepts has indeed only widened in scope to encompass contemporary molecular biology. We shall see, however, that holism contradicts the principle of the objectivity of nature on which scientific practice is founded, and that it reintroduces animism.

5.2 Philosophical holism

Without exhaustively reviewing the subject, we nevertheless want to highlight the points common to all versions of holism, and for this we shall base our remarks on the works of Conwy Lloyd Morgan (1852–1936), Samuel Alexander (1859–1938) and Jan Smuts (1870–1950). These three authors all played a major role in its development at the beginning of the 20th century. Smuts, moreover a major South African politician, seems to have been the first to use the word 'holism' in English in his book *Holism and Evolution* (Smuts, 1926). Morgan is also the author of a reference book entitled *Emergent Evolution* (Morgan, 1923).

Holism, which is summed up in the famous saying "The whole is more than the sum of the parts", is opposed to reductionism. It asserts that an entity possesses properties which can be neither explained nor predicted from the elements that make it up and that it thus forms an irreducible whole. According to this philosophy, when single elements enter into a relationship to create this whole, they are themselves altered by virtue of this relationship.

> "*A whole is a synthesis or unity of parts, so close that it affects the activities and interactions of these parts, impresses on them a special character and makes them different from what they would have been in a combination devoid of such unity or synthesis*" (HE p. 134).

The determining relationship between the single elements and the complex whole that they form is therefore defined in the concept of emergence. For the holist, while the whole has so-called resultant properties, which can be predicted from the properties of the elements, there also exist emergent properties which are not predictable. However, they cannot arise except from a particular material base corresponding to a specific configuration of and relationship between the elements. The concept of emergence refers both to what necessarily arises from a specific material base and to the non-predictability of what arises. It is important to note that this concept is deterministic. Each time the same base is produced the same emergence phenomenon is reproduced. As a first example let us look at the relationship between the properties of molecules and those of the atoms that make them up.

> "*When carbon having certain properties combines with sulphur having other properties there is formed, not a mere mixture but a new compound, some of the properties of which are quite different from those of either component. Now the weight of the compound is an additive resultant, the sum of the weights of the components; and this could be predicted before any molecule of carbon-bisulphide had been formed. One could say in advance that if carbon and sulphur shall be found to combine in any ascertainable proportions there will be such and such weight as resultant. But sundry other properties are constitutive emergents, which (it is claimed) could not be foretold in advance of any instance of such combination. Of course when one has learnt what emerges in this particular instance one may predict what will emerge in that instance under similar circumstances. One has learnt something of the natural plan of emergent evolution*" (EE p. 3).

This is a central idea of holism that is found among all its adherents:

> "*A mere mechanical aggregate is nothing new, and is no more than the sum of the mixed ingredients, while the chemical compound is new in the sense that out of the constituent materials another qualitatively different substance has been made. A new structure has been formed in the chemical compound. In the same way a new structure and substance is made in the atom out of the qualitatively different*

electrons and protons. It was on this account and in this sense that we called matter creative. Creative, that is to say, of structures and substances different from their constituent elements or parts. It is, however, when we come to consider organisms that we see the creative whole in a full and proper sense." (HE pp. 140–141).

In these emergent phenomena, holists see a further principle which has a fundamental place in their philosophy. They consider that matter is active and not inert. In their opinion, as proof of this, this creative activity, which can already be detected at the chemical level, is manifested with overwhelming evidence in the living organism. Again, as Smuts says:

"*An organism, like a plant or animal, is a natural whole. It is self-acting and self-moving. Its principle of movement or action is not external to itself but internal. It is not actuated or moved by some external principle of force, like a machine or an artificial construction. The source of its activity is internal and of a piece with itself, is indeed itself. It consists of parts but its parts are not merely put together. Their togetherness is not mechanical, but rests on a different basis. The organism consists of parts, but it is more than the sum of its parts, and if these parts are taken to pieces the organism is destroyed and cannot be reconstituted by again putting together the severed parts*" (HE p. 111).

All holistic philosophies and all the biological theories which are derived from them share this principle of attributing creative activity to matter. They discern evolution in nature characterised by successive emergences of totalities constituting qualitatively different levels of organisation of increasing complexity. The emergence of these levels from previously single elements is not reduced in this conception to a simple cause and effect relationship, nor is the whole compound the mechanical result of adding its elements together. It is a creation of something radically new. Holists in general distinguish three main levels emerging from this evolution. The first comprises matter, itself organised into sub-levels (macromolecules,

molecules, atoms and sub-atomic particles). At the second level, life emerges from the matter, and there are sub-levels here, too, which are superimposed in increasing complexity (animals, plants, multicellular and unicellular organisms). At the third level finally, emerges the mind, which is only present in Man. All holists support this general outline explicitly or implicitly, although with many variations.

Certainly, in explaining emergent evolution, the majority of them say that they reject the idea of a supernatural power separated from matter, such as a God, or vital force that transcends the laws of physics and chemistry. Yet while they deny the existence of such a power, all they are really doing is moving it elsewhere, endowing matter itself with this creative activity.

> "*The naturalistic contention*[24] *is that, on the evidence, not only atoms and molecules, but organisms and minds are susceptible of treatment by scientific methods fundamentally of like kind; that all belong to one tissue of events; and that all exemplify one foundational plan. In other words the position is that, in a philosophy based on the procedure sanctioned by progress in scientific research and thought,* ***the advent of novelty of any kind is loyally to be accepted wherever it is found, without invoking any extra-natural Power (Force, Entelechy, Elan, or God) through the efficient Activity of which the observed facts may be explained***"[25] (EE p. 2).

The efficient Activity which Morgan speaks of here has various names according to different authors. Alexander (1920) calls it 'nisus'. For Smuts, "*Holism is the term here coined for this fundamental factor operative towards the creation of wholes in the universe*" (HE p. 94). Earlier, Henri Bergson (1859–1941) had called it 'vital elan' (Bergson, 1907). It is however in every instance the characteristic mark of these philosophies, including when they are applied to biology where the creative factor has come to be called 'self-organisation'.

---

[24] As opposed to that of mechanism.

[25] Original text not in bold.

We are bound to acknowledge therefore that holism challenges the principle of the objectivity of nature which implies that matter is inert. Through reintroducing the idea of creative activity it once again dons a principle of internal movement, which is a characteristic of animism. The majority of holists may indeed reject the idea of a supernatural God, but they do so in order to immerse Him deeper in the very heart of nature. They may not support the idea of Creation as a separate original action, but they do not do so in order to reintroduce it more effectively in the form of continuous creation. For them, God is no longer external to the world He is creating but is immersed in it, and emergence is just the manifestation of His presence.

We shall now investigate whether these theories could help resolve the problem posed by the non-specificity of molecules, but first, several points that will facilitate this investigation need to be discussed.

Emergence can be understood as having a weak or a strong meaning. It indicates the creation of totalities which have non-predictable properties. However, this non-predictability could be subjective if it only depended on the imperfections of our cognitive capacity and not on the appearance in nature of properties which are really irreducible. For example, we cannot analyse certain processes because they involve too many parameters, so we say their properties are emergent. In this case, the concept of emergence only highlights the limits of our knowledge. It may be that with the development of research we will be able to go beyond those limits and that in the future, we may be capable of predicting and explaining these properties that have previously been considered irreducible. If that were the case, there would be no creation of radically new properties but simply an effect difficult to predict owing to the multiplicity of causes. This weak meaning of the concept of emergence presents no problem because it does not call into question the classic cause and effect relationship. For the holist, however, emergence means the creation of radical novelty (a structure or a property) corresponding to an objective reality in nature. With this strong meaning, it is a question of real creation *ex nihilo* and not of transformation of something that already exists in another form. Such a conception is of course irrational and incompatible with the scientific approach, which, in contrast, is based on reason. It could be justified as a religious argument but not as a scientific theory. In this respect, our intention is not to evaluate its validity, but to emphasise its extra-scientific nature. Holism's monopolising of the word evolution is just as problematical because it induces confusion with the theory of natural selection which is the utter antithesis of holism. For Darwinism, the organism is subjected to random variations which do not spontaneously create order and do not themselves end in producing new species. To do that, the selective action of the environment is required (see Fig. 4). The organism only acquires its structure under the influence of this external pressure and not from any internal trend which might give direction to its destiny. The theory of natural selection is therefore the product of a philosophy similar to that producing the principle of inertia. In its view the organism carries no internal determination and is not active in itself. It is constructed 'here and now' on being confronted with the world. It therefore completely opposes the theory of emergent evolution.

A final point deserves to be highlighted. Since genetic determinism is reductionist, holism would at first sight seem to be incompatible with it. Nevertheless, the two concepts unite in affirming the objective reality of order. In both cases a first principle is involved which structures the world and directs processes. In genetic determinism, the principle of order from order comes into play through the stereospecificity of the molecules (chapter 3), while in holism, the creative principle, less well defined and with a variety of names, creates organised wholes. Order is perfectly real in both theories, for the principle of order is inherent in matter. The wholes and their method of organisation exist objectively and not by virtue of any subjective divisions that our viewing of nature might operate on it.

For this reason, genetic determinism and holism agree on a single hierarchical model of the world consisting of the superimposition of levels of organisation of increasing complexity (Fig. 11).

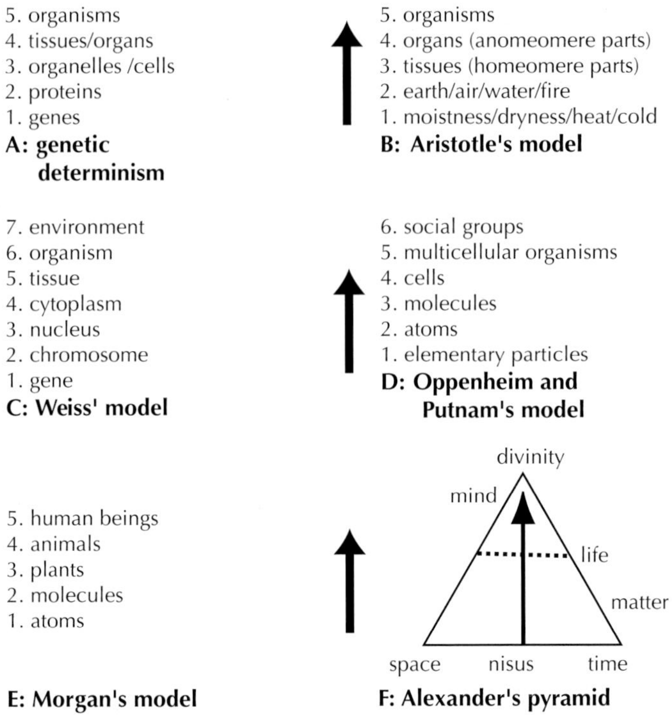

FIGURE 11. Indetermination of the layered model of the world. Six variations of the model are shown from a huge number of illustrations. **A:** Genetic determinism according to the molecular biologist Monod (1970). **B:** Ontogenesis according to the philosopher biologist Aristotle (see chapter 7 §7.2). **C:** The model of the embryologist Weiss (1973). **D:** The model of the philosophers Oppenheim and Putnam (according to Kim, 2002). **E:** The model of the philosopher psychologist Morgan (according to Kim, 2002). **F:** The model of the philosopher Alexander (1920). The first three of these models concern biology and the other three are general systems. They show all the differences reflecting their authors' specialities and research subjects. If the layered model of the world is of objective value, only one should exist. Which should be chosen?

For both theories the levels express universal typological realities corresponding to structures and fundamental modes of organisation. They are, in a way, a framework on which the world rests and out of which the diversity of individual things blossoms forth. The universal Molecule forms its own identity in concrete terms by becoming one of a diverse variety of particular molecules (water, carbon disulphide etc.). In the same way, universal types of Cells and Organisms differentiate into numerous individual identities: muscle, bone or blood cells, rabbits, carrots, foxes etc.

Both these conceptions are ontologies of order from order. They only differ in the origin of that order and its mode of production. For holism, order comes from the whole that is imposed on the parts: it is 'order from above' which reflects the "natural plan of emergent evolution" of which Morgan speaks (EE p. 3). For genetic determinism it is the reverse, order coming from the molecules that form the organism: this is 'order from below' which gives substance to the genetic information. In both theories however, order is always at the origin of order.

This vision of a world organised into superimposed levels where each thing has a specific place is nowadays hegemonic and seems absolutely obvious to us. Nevertheless, the philosopher Jaegwon Kim, who calls it the 'layered model of the world', has analysed it in detail, and thus revealed its fragility (Kim, 2002). Even though a great many researchers agree on it in principle, there are just as many variants of this model which do not acknowledge exactly the same levels. The differences between them, related to the historical context and the discipline in which they have been produced, raise the question of the objectivity of the layered model. Indeed, if it really exists, there must only be one, in which each level is generated from the preceding level and where each thing must be able to be placed in a unique position. The different variants would then be only approximations. Now, Kim shows that if we go into detail, the ideal is far from being realised. On the one hand, reality is often arborescent rather than layered. Animals are not superior to plants, as Morgan supposes, even though they manifest properties related to their having a nervous system. They are two separate evolutionary branches. Some classifications recognise a human level above that of animals, related to their having a mind or conscience. Others, such as that of the reductionist philosophers Paul Oppenheim (1885–1977) and Hilary Putnam, go directly from multicellular living beings to social groups. These two classifications are not compatible since there are social groups which are not human (insects, primates etc.). From the moment we recognise the human level, social groups can no longer be placed above it. If we place them below it, that means that the human emerged from a property common to all social groups (ants and primates for example). On the other hand, owing to their composition or properties, many organisms cannot occupy a sole level. Mammals are composed of cells and circulating molecules. They therefore arise from both these levels. Above which of them should we place mammals in the hierarchy? The blood system is a collection of cells and circulating molecules that ensure immune, nutritive, endocrine or respiratory functions. Should it be considered as cells, tissues, an organ, or as several organs? What about viruses? Are they living or non-living beings? Where should they be classified? Syncitia are multinucleated cells arising from the fusion of several cells: should they be put with cells or with multicellular organisms? Certain inanimate objects such as computers and robots are capable of remembering and calculating, even of demonstrating intelligence, characteristics which are usually considered as indicating a brain. Where are they to be placed in the layered model? These examples are only a minute sample from the multitude of problems that are encountered when the layered model is confronted with concrete cases. They just go on to show that we can legitimately doubt the ontological reality of the layered model. Far from representing the intrinsic organisation of the world, its different variants seem rather to indicate a mode of subjectively dividing up reality depending on the observer. Noble arrived at a very similar conclusion from his work on cardiac physiology, and has put forward what he calls a theory of biological relativity. Without formally denying the existence of levels of organisation, he thinks that none of them has any privileged causal role. They can all be used as a starting point for analysing the living organism (Noble, 2006, 2008), which suggests that they are of epistemological rather than ontological value.

5.3 Biological holism

The different kinds of biological holism include neo-vitalistic theories and theories of self-organisation. Neo-vitalistic theories are not really fashionable any longer but their authors have often raised pertinent questions, even if the answers they provided are poor. We will therefore tackle these two types of holism in succession. Our review will not be exhaustive, any more than it was for philosophical holism, but we shall try to define their general characteristics in order to assess their ability to resolve the problem of molecular non-specificity.

5.3.1 *The neo-vitalistic holism of Hans Driesch*

Hans Driesch (1867–1941), one of the pioneers of experimental embryology, supported a vitalistic theory (Driesch, 1908, 1914). Another pioneer in embryology, Roux, had, in an experiment on the frog, destroyed one of the two cells of the embryo after the first division, thus succeeding in inducing the development of half an embryo. This seemed to confirm the theory of August Weismann (1834–1914), the forerunner of genetic determinism, which postulated the existence of a highly organised microscopic material structure in the germinal cells that he called 'germinative plasma'. This structure, which foreshadowed DNA, was thought to control embryonic development in a very precise way, with each of its parts determining a part of the adult organism. Weismann also imagined that the germinative plasma was split at each division of the cell. As each embryonic cell only received a portion, it could therefore only form one specific region of the adult organism[26] corresponding to the portion of germinative plasma that it had received. This theory easily explained Roux's experiment. The surviving cell must have contained only half of the germinative plasma relating to half the organism.

[26] Unlike DNA, which is present in its entirety in each cell of the organism.

Driesch performed an experiment similar to Roux's on a sea urchin, but he obtained a completely different result. Instead of inducing the formation of half an embryo, destroying one of the two cells of the embryo ended in the formation of a complete animal, though of a reduced size. Driesch repeated his experiment on embryos at the four cell stage and was able to show that either a single cell or a group of three cells taken together is capable of forming a complete embryo. In the same way, half an older embryo containing a thousand cells can produce a normal organism. These results invalidated Weismann's theory. If cells resulting from several successive divisions can form complete organisms, they must contain all the germinative plasma, not just some of it. Driesch drew an additional conclusion from these experiments, which demonstrate that up to an advanced stage, one cell of an embryo can give rise to all the cell lines of an organism, and that it possesses a potential for differentiation greater than its actual vocation during embryogenesis. By multiplying experiments on embryos taken at different stages of development in a variety of experimental conditions, he was able to verify this conclusion and demonstrate the plasticity of cells, which enables them to adapt to these varied situations. The question then arose of how this immense potential for development is reduced during embryogenesis so that only a single potentiality is expressed.

To answer this, Driesch performed other experiments which led him to formulate two further concepts. Firstly, the development potential of all the cells from one region of the embryo seems to be constant. The embryo is therefore 'an equipotential system'. Secondly, to a certain degree, one region can be modified without interfering with the development of other regions of the embryo. The development of the various regions seems causally independent, but nevertheless it ends in a harmonious organism forming an organised whole. Driesch described this phenomenon as 'a harmonious equipotential system'. In fact what he was describing corresponds to essential properties which are nowadays completely acknowledged, and which we now call cellular plasticity and robustness of biological systems.

These properties cannot be compared with those of a simple machine. How can we account for this? The laws of physics and chemistry and mechanistic determinism are not sufficient to do so, in Driesch's opinion, and he evokes the action of a vital force that he calls 'entelechy' (Driesch, 1908, 1914), but such an explanation is obviously not acceptable to us. We do have to acknowledge however that giving prominence to cellular plasticity is still relevant and the issues raised by Driesch are still topical. They are, indeed, similar to those posed by the non-specificity of proteins. In both cases the potential for differentiation of the cells, or for interaction between molecules, is greater observed *ex vivo* than *in vivo*. Without adopting Driesch's vitalism, it is necessary to take this into account when explaining ontogenesis.

5.3.2 *The neo-vitalistic holism of Walter Elsasser*

Molecular biology has been influenced by the work of Delbrück and Schrödinger, while another physicist, Walter Elsasser (1904–1991), was also interested in biology, but from a completely different point of view, for he put forward a vitalistic conception (Elsasser, 1998). According to Elsasser, the living organism is so complex that it cannot be analysed with the laws of physics and chemistry. He based his proposition on a number of arguments. If the 10^{12} atoms of a cell were solely controlled by the laws of physics it would be possible to calculate how many combinations of them would be possible. This number is greater than 10^{100}, i.e. far greater than the number of protons in the universe, namely 10^{80}. There is no corresponding physical reality or operational value. One can also calculate the total number of a given type of cells that exist or have existed in the history of the Earth. This is in the order of 10^{42}. The number of possible molecular states is therefore hugely in excess of the number of cells actually existing. On account of this it might reasonably be thought that the living organism is highly undetermined at molecular level. Elsasser started from this hypothesis for his theoretical elaboration. Initially, he acknowledged that it cannot be proved immediately, but only its plausibility progressively demonstrated.

To do this, he focused on the heterogeneity of individuals in the biological classes. If a small part of a body or of a cell is defined and compared with an equivalent part from another individual of the same class, they are always different. Whether it is at the anatomical or biochemical level, this heterogeneity is such that a description of an average individual, characteristic of a type (cell or species), is completely false. There is radical interindividual heterogeneity in biological classes and this heterogeneity confirms the hypothesis of the microscopic indetermination of living beings.

Such microscopic indetermination puts paid to the idea that order in the living organism may result from molecular order. According to Elsasser, it is necessary to formulate a new biological theory in order to explain that biological organisation can be maintained despite this microscopic indeterminism. He suggested four principles to define this theory.

He called the first 'ordered heterogeneity'. In biological systems where there is large scale regularity, there is small scale heterogeneity. Order is maintained for the whole despite heterogeneity among the parts. This is a principle obviously different therefore from both that of order from order, of molecular biology, and from that of order from disorder, of statistical physics. Indeed, according to the principle of ordered heterogeneity, order is not the result of reducing microscopic variability by the law of large numbers but is a real property of the macrostructures (the wholes), and not a mathematical approximation, as in statistical physics. It is identical to the principle of macro-determinism suggested by certain biologists (see this chapter, §5.3.6).

The second principle is 'creative selection'. During ontogenesis, the number of molecular configurations is restricted as the organism selects a certain number of them from among the huge range of possibilities. This principle attributes real creative properties to living matter, and despite using the term 'selection', it is totally the reverse of Darwinian selection.

The third principle is 'holistic memory', and is supposed to explain how the first two principles function: information relating to the overall structure of the organism is stable and transmitted directly from generation to generation without a material storage mechanism as is the case with the genetic information encoded in the DNA.

The last principle is 'operative symbolism'. Elsasser does not deny genetic heredity, but it must be completed by holistic memory. Without going into detail, this operative symbolism is said to control the relationships between the two types of heredity, genetic and holistic.

Thus, like Driesch, Elsasser managed to establish irrational principles which transcend the normal context of scientific logic and approach. In both cases, the explanations attribute powers to living organisms which exceed the laws of physics and chemistry. Driesch's entelechy along with Elsasser's four principles are concepts which go no further than a simple linguistic, virtually magic, formula: they have no real content and are consequently incapable of explaining in concrete terms how biological systems are organised. However, it must be recognised that Elsasser's ideas are based on pertinent elements of analysis which still have a firm hold in current debates. Like the plasticity and robustness brought to the fore by Driesch, the enormous number of possible molecular combinations in living systems that Elsasser emphasises is an effective property generated by proteins. We have to include it in our explanations, while providing a rational answer to the problem it raises. The mistaken wanderings of Driesch and Elsasser reveal how important it is to respond to this problem in order to avoid the irrational excesses of holism.

5.3.3 *Self-organisation according to Prigogine*

There are a great many self-organisation theories, and in what follows in this chapter, we shall illustrate the main variations which are based on physics, cybernetics and biology.

The physicist Ilya Prigogine (1917–2003) was very influential in this area. He studied so-called open systems that receive energy or matter from their environment. When these systems are subjected to fluctuations, instead of returning to their initial state they may evolve towards another more ordered state. In so far as fluctuations are inherent in any physical and chemical system (see chapter 2, §2.2.4), this phenomenon has been likened to self-organisation. The system is supposed to be capable of creating order spontaneously without being influenced by an external cause. Prigogine called these types of system "dissipative structures" in order to indicate that the creation of order is accompanied by dissipation of energy. The concept is used now to describe a large number of processes. Adherents of self-organisation consider the origin and functioning of living beings as relating to dissipative structures. We can quote the two main examples of this given by Prigogine in his book *Order Out of Chaos* written in collaboration with Isabelle Stengers.

The appearance of the ordered movement of molecules forming convection cells in a heated fluid, initially observed by Bénard (1874–1939), is considered as the paradigm of self-organisation.

> "(...)*Bénard instability is another striking example of the instability of a stationary state giving rise to a phenomenon* ***of spontaneous self-organisation****. The instability is due to a* ***vertical temperature gradient***[27] *set up in a horizontal liquid layer. The lower surface of the latter is heated to a given temperature which is higher than that of the upper surface. As a result of these boundary conditions, a permanent heat flux is set up, moving from the bottom to the top. When the* ***imposed gradient***[27] *reaches a threshold value, the fluid's state of rest, the stationary state in which heat is conveyed by conduction alone, without convection, becomes unstable.* (...) *convection corresponding to the coherent motion of ensembles of molecules is produced, increasing the rate of heat transfer. Therefore, for given values of* ***the constraints (the gradient of temperature)***,[27] *the entropy production of the system is increased; this contrasts with the theorem of minimum entropy.* (...)*Bénard instability is a spectacular phenomenon. The convection motion produced actually consists of the complex spatial organisation of*

[27] Original text not in bold.

the system. Millions of molecules move coherently, forming hexagonal convection cells of a characteristic size" (OOC p. 142).

Another example of a dissipative mechanism given by Prigogine concerns the creation of concentration gradients of morphogen molecules during embryogenesis. This is a system of coupling between several chains of chemical reactions in which the various products of the reactions diffuse with different speeds (OOC pp. 146–153). In this system the reactions are specific and the concentrations of the reactants are fixed by interaction with the medium. For certain values of these concentrations, instead of achieving a classic state of chemical equilibrium characterised by constant concentrations of the reaction products the system oscillates in cycles, with the concentrations varying reproducibly over time and space. Prigogine's model is in fact similar to Alan Turing's reaction-diffusion mechanism.

In Prigogine's view, these phenomena provide support for a holistic philosophy, and in his view, "*This leads to a new view of matter in which matter is no longer the passive substance described in the mechanistic world view but is associated with spontaneous activity*" (OOC pp. 9). The spontaneous emergence of order in dissipative structures bears witness to the creative trend that animates nature. This is the ordinary holistic principle which is a central theme of his theory.

Dissipative structures are therefore supposed to explain the creation of order in living organisms. Can they however explain how non-specific biological molecules are organised during ontogenesis? To answer this question several points must be borne in mind.

First of all, the coupled chemical reaction model cannot by definition be relevant, because the chemical reactions brought into play are specific. What we need is a model which explains the appearance of order from numerous molecules involved in a great many non-specific interactions (and reactions), comparable with biological systems (see chapter 4). How does Bénard's instability come into this? It is always given as the very prototype of self-organisation demonstrating that life is thermodynamically possible.

Now while there is absolutely no doubt about Prigogine's description and mathematical modelling of the phenomenon, its conceptualisation and generalisation are very problematical. The experimental phenomenon, that is to say the formation of convection cells, has to be clearly distinguished here from the theorisation about it in the context of self-organisation. As Prigogine himself explains, the phenomenon of organisation depends on a temperature gradient which is an external constraint imposed on the system. Consequently, "*Bénard cells, like all dissipative structures, are essentially a reflection of the* ***global situation***[28] *of non equilibrium producing them*" (OOC pp. 143–144). We must insist on this point, as it is important and leads many researchers astray: the external global constraint involved here belongs to the reality of the phenomenon. Is it right in these conditions to use the concept of self-organisation to describe this reality? There is a flagrant contradiction here between the phenomenon described and its conceptualisation. It would be more exact to speak of hetero-organisation to indicate the fact that the system is organised under the effect of the constraint arising from the environment. The organisation produced depends on this constraint and not on a phenomenon of spontaneous emergence from the components of the system. Bénard's instability is no exception. Other biological systems alleged to be self-organising are in actual fact determined by constraints, as is particularly the case of the organisation of colonies of social insects, which is often given as an example but which depends in reality on environmental factors, especially the substrate sources which feed the colony (Camazine *et al.*, 2003).

Finally, theories of self-organisation are subject to another confusion that needs to be elucidated. They are often assimilated into probabilistic theories. Now while there is indeed a random event, fluctuation, in dissipative processes, which introduces a degree of uncertainty, it is only involved as an event triggering a deterministic dynamic. Dissipative mechanisms are not therefore intrinsically probabilistic but are deterministic with noise (see chapter 2, §2.2.4).

[28] Original text not in bold.

5.3.4 *Self-organisation according to Stuart Kauffman*

Stuart Kauffman is another eminent self-organisation theorist, who, like Prigogine, advances the themes of holism. For Jacques Monod, life appearing was an event that had virtually no chance of occurring (CN pp. 131–137) and we are therefore the result of an accident of history, strangers in the world we inhabit. Kauffman (1993, 1995) categorically rejects this point of view, believing that the appearance of life expresses a phenomenon of spontaneous emergence that was inevitable. This is why in contrast, we are *At Home in the Universe* as the title of one of his books says. In addition, the living world exhibits an order which cannot, in his opinion, be explained simply by natural selection. He asserts that the main organising force is a spontaneous trend towards self-organisation (HU pp. 23–30) and backs his views with work performed with Boolean automaton networks.

A Boolean automaton is an entity (an electric lamp, an enzyme, a gene etc.) which can be activated or repressed. A numerical variable describes its state (1 or 0 respectively). In a network of a succession of Boolean automata, the state of each of them depends on the state of the others. The Boolean networks studied by Kauffman to support his conception of self-organisation are deterministic. The state of each node of the network (each automaton or entity) depends on the state of the nodes situated upstream, according to Boolean rules using the operators AND, OR, and EXCEPT. For example if a node x depends on the state of three nodes a, b, c, a rule could be: x is active if a AND b AND c are active. Another rule would be: x is active if one of the nodes a OR b is active, etc.

The first experiment concerns the problem of the origin of life (HU pp. 54–66). Kauffman considers a living system as a huge Boolean automaton network, each automaton representing a protein or a gene. If, bathing in the primitive soup where life germinated, there were thousands of chemical components of the first living network, what was the probability of it arising simply through molecular encounters? Kauffman showed in a computer simulation that this probability depends on the connectivity of the network, i.e. on the average number of potential connections for each node.

To connect two nodes of a network a link is needed with a ratio of 0.5. From the moment this value is exceeded, i.e. when the components of the network can be connected by multiple links, the probability of forming a single network connecting all the components is greatly increased until this event becomes inevitable. Kauffman assumes that this is how life occurred. It was not therefore an accident but the result of this spontaneous tendency to self-organize shown by high connectivity Boolean automata. This first result seems compatible with the structure of real protein networks the nodes of which are indeed highly interconnected (chapter 4, §4.1.5). However, this experiment does not take into account the fact that biochemical networks also have to be functional. It does not indicate whether the networks formed are ordered or chaotic. Kauffman investigated this question in another study dealing with cell differentiation (HU pp. 71–112).

A differentiated cell results from the stable expression of a sub-set of genes of all those forming a genome. The question is therefore whether Boolean automaton networks generate ordered states in which the same nodes (genes) are constantly (or cyclically) activated. Kauffman's results show that this is possible on condition that each node at the most only depends on two other nodes of the network. If the connectivity of the network is greater, the latter very rapidly becomes chaotic. This is a problem in itself in regard to experimental reality, since the connectivity of actual networks is very great. However, a high connectivity network could still produce ordered states if it were biased so as to direct its behaviour towards stable states. Gérard Weissbuch, who perfected the model, calls this bias the p parameter (HU pp. 84 and 103; Weissbuch, 1999). Put simply, it means that, as in the case of Bénard's instability, a constraint needs to be exerted on the Boolean automaton network for it to organise itself. Although it does not seem to have occurred to the adherents of self-organisation, this ruins what Kauffman wants to demonstrate, since it is not spontaneous organisation.

Do these models help resolve the contradiction in genetic determinism caused by the non-specificity of molecules? When Kauffman carried out his work, no actual networks of proteins or genes had yet been thoroughly studied. We now know that the connectivity of these networks is very high. According to Kauffman's own results, such networks cannot follow Boolean deterministic rules because they would be totally chaotic. Kauffman's Boolean networks, which are based on the Monod-Jacob model of stereospecific regulation (see chapter 3, §3.3), do not therefore help in explaining how actual networks of non-specific molecules function. From a theoretical point of view, they lead to the same paradox as Prigogine's dissipative structures. They have to include the action of a constraint arbitrarily applied on the system, and thus contradict the very idea of self-organisation.

5.3.5 *Self-organisation according to Atlan*

Cybernetics has also given rise to a theory of self-organisation (Segal, 2003). Disturbances occur in any communication channel and affect the signal being carried. These disturbances, called "noise", usually have a negative effect, for example, in a television network where the image may be fuzzy or the sound of a telephone link inaudible. Several researchers have suggested that, instead of exerting a negative effect, noise may have a positive role in allowing a perturbed system to self-organise.

Heinz von Foerster (1911–2002) suggested a principle of "order from noise" (von Foerster, 1960). As in Turing's reaction-diffusion system (chapter 2, §2.2.4), if noise were to affect a system sufficiently and intensely, the system could depart from its state of equilibrium and evolve towards another more complex state. To illustrate this principle, von Foerster used the image of a formless set of magnets. If the set is shaken about, the magnetised surfaces of the parallelepipeds will stick to each other and the mass will be transformed into a more complex shape (Fig. 12). In this procedure, the noise (shaking the magnets) causes self-organisation of the system.

In the wake of von Foerster, Henri Atlan has also formulated a theory of self-organisation from noise (Atlan, 1972, 1979, 1999). He has demonstrated that if some of the elements of a system are

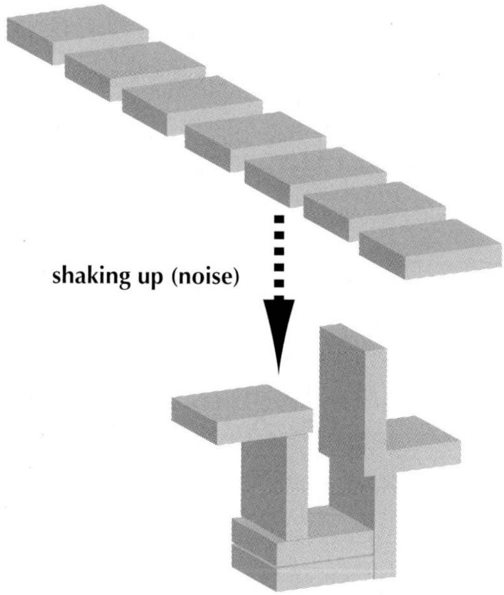

FIGURE 12. Von Foerster's magnets. Shaking up the magnets increases the complexity of the structure, which is equivalent to a self-organisation process.

redundant, noise will increase the quantity of information in the system rather than decrease it. This increase will induce it to self-organise due to the increase in complexity. The principle of this theory is simple to understand. If a system is composed of several identical entities, noise randomly modifies each entity and thus creates a variety of structures richer than the initial homogeneous set. If each element corresponds to a piece of information, e.g. to a gene, the total quantity of information likewise increases.

> *"In the transmission of information between DNA nucleotide sequences and protein amino-acid sequences for example, it is known that there are always errors equivalent to what is called noise in a communication channel. It is easy to conceive of these errors producing a negative effect which, in the formalisation of quantities of information, developed by Shannon, results in a quantity of the information transmitted being deducted. The effect of*

noise is a reduction in the information carried by the protein relative to what it would have been if the transmission were perfect, i.e. if the protein strictly corresponded to the DNA. If, however, instead of considering the transmission of the information from its source to its arrival, one were now to envisage the total quantity of information in the entire system, of which this transmission pathway is but a part, one can quite easily show that the quantity of information produced by the noise is added and not subtracted. That can be understood intuitively: errors end in a protein, the structure of which is not an identical reproduction of that of the DNA, and they therefore introduce new variability which represents diversity, compared with what would happen if there were no errors. This diversity can obviously be the source of poor functioning and produce negative effects, but in certain cases, it may on the other hand be the source of an increase in complexity, and possibly of functional complexity — with an overall positive effect for the system"[29] (FTG pp. 25–27).

In order to translate this theory into a concrete biological mechanism, which is capable of explaining how cells function, Atlan stressed the sources of redundancy in organisms. He particularly emphasised the existence of DNA sequences which are repeated countless times in the genome of multicellular organisms, his idea underlying this being that their mutation during embryonic development could play a functional role, as is the case with the synthesis of antibodies in the immune system, (CF pp. 70–72; FTG p. 28). In addition, there are bound to be random variations in the concentrations of biological molecules inside cellular compartments, caused by thermal agitation. Atlan also suggested that these variations may be a source of self-organising noise.

However, as he himself emphasised, whether the noise comes from the environment or has to do with fluctuations in chemistry or diffusion, it is always a disturbance to normal functioning of the system and as such, is an external factor (CF pp. 56–57, 81–82).

[29] Translated from the French by Margaret Hutchings.

The mechanisms activated in his theory are therefore deterministic mechanisms with noise. This is particularly well illustrated by von Foerster's example of the magnets from which he took his inspiration. The laws of the interaction of magnets are deterministic. Depending on their polarity, magnetised surfaces attract or repel each other. Agitating the magnets (the noise) only serves to set off a new dynamic which will end in another state of organisation. For Atlan, similarly, in actual biological systems, genetic information is still an essential notion and proteins are always bearers of information arising from their three-dimensional structure (FTG pp. 25, 33). Noise only intervenes through perturbing their effects and modifying the way the networks they constitute function. Like other self-organisation theorists, Atlan has never questioned stereospecificity and the deterministic functioning of genes, which ensues from it. His theory is still situated therefore in the context of the principle of order from order, and as a result, his views conform to conventional theories of embryogenesis. He believes that Turing's reaction-diffusion mechanisms create the gradients of morphogenic molecules that control genetic expression (FTG pp. 44–47).

5.3.6 *Self-organisation according to Weiss*

Renowned experimental biologists have also for a long time been providing support for the idea of self-organisation based on their own work, among them the embryologist Paul Weiss (1898–1989), who played a prominent role. For Weiss, the organism is not constructed from the gene, but is produced from the multiplicity of interactions between the various levels of organisation going in both directions, from the organism to the gene and from the gene to the organism (Weiss, 1973). Two forms of determinism may exist: macro-determinism corresponding to descending causality (from the organism towards the gene) and micro-determinism corresponding to ascending causality (from the gene towards the organism). Macro-determinism is said to predominate over micro-determinism (SL pp. 10–13).

Weiss supported his theory with empirical observations. Territories in an embryo can be defined in the knowledge of exactly what they will become at a later stage of its development. In any embryo the same territory will become the same part, but there is a lack of determination as to what the parts (the cells) inside this territory will become (SL pp. 21–22). A particular part in one embryo does not have exactly the same future inside the territory as that same part in another embryo. This phenomenon would be found at every level of organisation, e.g. in cells as regards their molecular constituents. Cells of the same type are globally identical but there is microscopic variability between them. For Weiss, these observations reveal a general principle "*... of determinacy in the gross despite demonstrable indeterminacy in the small for practically any level and area of the life sciences*" (SL pp. 21–22). Macrodeterminism of the global structure of the organism would progressively constrain the lower levels without their being totally determined in detail (SL pp. 23–24). In this process, the macrodeterminism would not be reduced to the properties of the parts of the organism (the tissues, cells, molecules): on the contrary, the macroscopic structures and properties would emerge during a self-organisation process (SL pp. 29–35).

To explain what he means by self-organisation, Weiss uses the photograph of a vast beach taken one sunny Sunday when there were a lot of people bathing there (Fig.13). The beach is marked at the top of the photo by the edge of the sea which is slightly wavy and at the bottom by a straight road. The people look like dots which are denser near the water's edge and in various places corresponding to restaurants or different attractions. If the same photo were taken on another Sunday that was just as sunny, the picture would appear to be identical although in detail each dot (each bather) would be different and would not be strictly in the same place. The same structure would therefore be produced although the individual behaviour of the bathers would be different. According to Weiss, this is a phenomenon of self-organisation. It is characterised by the emergence of a type of collective behaviour

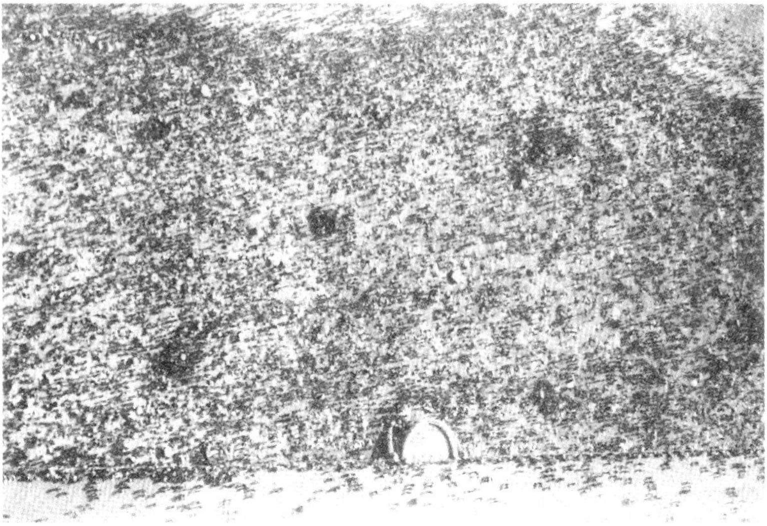

FIGURE 13. Self-organisation of a beach according to Weiss. Every time this beach is photographed on a sunny Sunday during the same season, with an identical number of visitors, an overall identical picture will be obtained although the details vary. We are grateful to Blackwell Publishing for permission to reproduce this figure (Weiss, 1973).

among bathers which is produced without their individual behaviour being coordinated by any precise determinism.

> "*Consider the people as molecules. The heavier border on top is the condensed belt of hydrophobic bodies adsorbed to the water-beach interface. The dark clusters inside the mass clearly mark domains of attractive forces, presumably emanating from sources of nutrient and stimulant attractants. Their equidistant spacing indicates mutual repulsion through forces of competition; and so forth. The analogy is not at all facetious. It cuts deep into the heart of our topic, for it exemplifies basic features of self-organising systems*" (SL p. 30).

According to Weiss, these self-organising properties of the living organism imply that the reductionist point of view should be abandoned in favour of the holistic view. Living beings make up systems with irreducible properties that he explains more precisely using the field concept.[30] Like particles which conform to the magnetic field in which they are placed, each individual cell of an embryo, or tissue, must conform to the morphogenetic field to which it belongs.

> "*Let us take a circumscribed body, depending for its maintenance on active exchange with its environment; for instance, an egg in a pond, a cell in a tissue, a human individual in a society. Then let the unit multiply into a few more units; they all continue to have a share in the common interface of exchange and communication with the medium. But let the number of units keep on increasing, whether by subdivision or accretion, and all of a sudden a critical stage arises at which some of the units find themselves abruptly crowded inward, cut off completely from direct contact with their former vital environment by an outer layer of their fellows. The latter thereby acquire positions not only geometrically, but functionally mediatory, between the ambient medium and the now inner units. From then on, 'inner' and 'outer' units are no longer alike. A monotonic group of equals has become dichotomised into unequal sets. With the emergence of the distinction between innerness and outerness, the* 1 + 1 = 2 *rule becomes inapplicable*" (SL pp. 31–32).

The morphogenetic field corresponds to this external/internal polarity which is propagated within a population of cells and causes their differentiation. Indeed:

> "*Interactions between the 'outer' members and their newly established inner' neighbors would expose to another set of new conditions any fresh units arising subsequently in the intermediate zone between them, and hence call forth in them a third type of reaction. Moreover, polarised influences from outside would impose an axiate pattern upon the group. Thus would ensue a train of sequelae of*

[30] Since Weiss, the idea of the morphogenetic field has been widely used in embryology.

ever-mounting, self-ordering complexity. In all these steps, the fate of a given unit would be determined by its response to the specific conditions prevailing at the site in which it has come to lie, those conditions varying locally as functions of the total configuration of the system — its 'field pattern', for short" (SL p. 32).

Does this version of self-organisation resolve the problem of the non-specificity of proteins better? Weiss' analysis is very interesting but like all the theories of self-organisation, his theory contains a contradiction: it includes the action of external constraints without it being explicitly accepted. In the example of the beach, the shape and size of the structure created depend on the position of the water's edge and the road. If these constraints were to change position or nature, not only would the general shape of the beach change but also the way it is structured internally. Depending whether the road is a small lane or a motorway with car parks arranged to allow access by huge crowds, the beach will be more or less visited and therefore there will be more or less restaurants. It is the same with cell populations. The morphogenetic field can only exist in as far as cells have a relationship with the environment which lays down a structuring polarity for the system. If the cells were independent of it, the field would have no reason to occur. These examples once again illustrate a phenomenon of hetero-organisation and not of self-organisation. Organisation does not emerge spontaneously from the local interactions of the basic constituents (bathers, cells) but ensues from the action of environmental constraints. Finally, like all adherents to self-organisation, Weiss never challenges the deterministic mode of functioning of genes and proteins. He simply thinks that their influence is delimited by the emergent properties of living beings (macro-determinism).

5.3.7 *Self-organisation according to Kirschner, Gerhardt and Mitchison*

Other biologists have tried to apply the concept of self-organisation to molecular biology, including Marc Kirschner, John Gerhardt and

Tim Mitchison. Their conception is based on the distinction between self-assembly and self-organisation (Kirschner *et al.*, 2000). We have seen that self-assembly is a process based on stereospecificity (chapter 3, §3.2). It leads to the formation of unique structures in stable equilibrium which do not require the exchange with the environment of any matter or energy. In contrast, self-organisation would imply consumption of energy and constant exchange of matter with the environment. In the first instance, it would produce dynamic equilibria between several states, then, owing to constraints or amplified random fluctuations, instead of oscillating between these states, the system would swing towards a specific final state.

To illustrate their theory, Kirschner *et al.* provide several examples. Certain bacteria produce a polarised tail-like bundle of filaments of actin proteins which helps them be propelled along. At the start the bundle is symmetrical but is subject to random fluctuations between which it oscillates. If one of them is too strong, it will be amplified and cause polarisation of the bundle. Another example concerns the differentiation of embryonic cells. The two essential elements for this process are signalling between cells and cell competence. Cell competence means that embryonic cells have several possible differentiation pathways at a given stage of their development. For Kirschner *et al.* this is a question of equilibrium between several interconvertible states, each state corresponding to the potential initiation of one of the differentiation pathways. Under the influence of signals from other cells, the equilibrium would be biased in favour of one of the pathways, which would then be selected.

Kirschner *et al.* insist on the dynamic aspects of ontogenesis and have integrated the role of random fluctuations. In this respect they go beyond conventional genetic determinism. However, like other adherents of self-organisation, they do not challenge it head on. They continue to accept the stereospecificity of interactions between molecules and genetic information and integrate this in their theory. They put forward a synthesis to explain how information is transmitted in biological systems. It is more complicated than the outline of conventional genetic determinism because it adds a stage of self-organisation to self-assembly based on stereospecificity, then swings towards a particular phenotype. However, it is not fundamentally different from it. It is still a sketch of construction of the organism which starts from the genes and works upwards to the phenotype. Genetic information determines the folding of the proteins which spontaneously self-assemble owing to their three-dimensional structure. Their theory again falls within the context of the principle of order from order, and cannot resolve the problem of the non-specificity of molecules.

5.4 Self-organisation does not exist

Very many biologists reduce self-organisation to a theory which seems to reject genetic determinism while acknowledging that biological organisation emerges spontaneously from local interactions between molecules (Camazine *et al.*, 2003). This widespread simplification leads to the worst confusions. It neglects what the concept of emergence really means. In truth, molecular biologists have never denied that an organism forms from interactions between molecules! They have formulated the concept of self-assembly to describe this. To them it is obvious (Britten, 1998), and Jacques Monod spoke in this respect of spontaneous morphogenesis (CN pp. 82–88). The ideas of emergence and self-organisation go a great deal further. They imply that there is creative activity in matter which makes emergence possible, from their single elements, of totalities with irreducible properties. Such a phenomenon would involve inexplicable creation and go beyond the rationality of science. A conception of this nature cannot resolve the problem posed by the non-specificity of molecules. Being content to assert that order emerges spontaneously from interactions between molecules without suggesting a mechanism for this falls into the realm of magic. The adherents of self-organisation have really tried to formulate more elaborate models to explain emergence, but all these models are based on stereospecific molecular interactions.

Like genetic determinism, self-organisation presupposes that order is real, and the emergence of levels of organisation of the world expresses a principle of order immanent in matter. It is therefore incapable of taking into account the non-specificity of biological molecules which demonstrate the opposite. When it tries to explain real phenomena, its application models contain environmental constraints, without their being explicitly conceptualised. Certainly, they may be relevant in as far as they describe an experimental phenomenon, as in the case of Bénard's instability model or Weiss' embryogenesis model, but they contradict the very idea of self-organisation which gave rise to them because in reality they are models of hetero-organisation. Not only do theories of self-organisation not resolve the contradiction in genetic determinism but analysis of them demonstrates the need to integrate environmental constraints to explain organisation of a system in a theoretical context which goes beyond both genetic determinism and holism.

6

Hetero-organisation

SUMMARY. The organism is the result of a process of hetero-organisation. Since molecules are non-specific, randomness is introduced into the interactions between proteins, generating very many possible structural combinations. This large number of combinations is useful to living organisms, as it produces the diversity of cells necessary for building the organism's tissues, and during ontogenesis, it is controlled by a selection process. Each cell adapts to its microenvironment made up of the other cells of the organism, this microenvironment itself depending on the external environment. Natural selection thus takes place in the internal environment and is the causal agent for formation of the organism. Ontogenesis and phylogenesis constitute a sole ontophylogenesis process which excludes all finalism. As Bernard suggested, it is the organisation of a multicellular living organism which ensures the life of the cells that constitute it, not the reverse. This conception resolves the contradiction in genetic determinism because a historical explanation replaces an explanation in terms of levels of organisation. The living organism is produced neither by the molecular level, nor by the cellular or organismic level, as is supposed by reductionism and holism. It is the result of its history. This theory is in line with a vast amount of experimental data which demonstrate that the differentiation of cells and gene expression are stochastic phenomena. It integrates the role of non-specific signals, though the latter are not inducing agents, only contributing towards the selection and stabilisation of cell types. Ontophylogenesis also helps us to understand the probabilistic functioning of the genome, the structure of which, like that of the cell, is the result of the evolutionary history of the organism. The genomic structure does not act like a deterministic genetic programme, but determines the probabilities of gene expression during embryogenesis. Computer simulation demonstrates that ontophylogenesis is endowed with the general properties expected of a theory of embryogenesis, particularly the creation of organised cell tissues, and also suggests that cancer should be understood in a new light. Biological organisation is produced by equilibrium between the selective constraints exerted on the organism and the stochastic character of the interactions between molecules, particularly those which induce gene expression. Cancer is produced as a result of imbalance between these two components of ontophylogenesis.

Genetic determinism is trapped by the contradiction within it. It has to call on cell structure to sort the interactions between non-specific molecules, but since, according to this theory, it is the genes that direct the way living organisms are constructed, the structure itself must be the product of these interactions. Holism provides no solution to this problem, other than evoking the emergence of cell structure *ex nihilo*. We must therefore pursue our search for another principle.

In solving a scientific problem one must not be afraid of drawing the most extreme consequences from experimental facts, even if they challenge the current theory. Its coherence is initially destroyed but later a new theory is devised which is even more relevant. The theoretical work consists in reinterpreting already known experimental facts in a new conceptual context. This is the path we shall follow now. We have to accept both the consequences of the non-specificity of the interactions between molecules and the role of cell structure, while going as far as possible with the analysis in order to unearth the significance of these experimental facts. For holism, the structure of the cell is a whole produced by emergence, but evolutionary biology sees it as the result of its history during which it was fashioned generation after generation by natural selection. If we take this further, it means that natural selection is involved, *via* cell structure, directly in ontogenesis, and not just in evolution. Natural selection shapes cell structure which, in turn, sorts interactions which are by themselves random, between non-specific molecules. This process of selection allows ontogenesis to occur.

Extending the scope of Darwinism in this way obviously upsets the current synthetic theory of evolution (Mayr and Provine, 1998; Huxley, 1942), where ontogenesis and evolution come under two different mechanisms. Natural selection is only involved in evolution *via* the mutation of genes which, in turn, determine ontogenesis. The hypothesis that we are advancing, a concept which clashes with our usual mode of thought, is that the two processes are produced by a single mechanism (Fig. 14). Instead of being the result of a deterministic mechanism controlled by the genes, ontogenesis is understood to be an intrinsically probabilistic process, as the stochastic interactions between molecules are subjected to selection by the cell structure, which is itself selected by the organism's environment.

In actual fact, the idea of ontogenesis resulting from selective Darwinian rules is not absolutely new. In Antiquity, Empedocles (490–435 B.C.) also resorted to a mixture of chance and selection to explain it.[31] In the 19th century, Roux wrote a book called *Der Kampf der Theile im Organismus* [*The Struggle of Parts in the Organism*] (1881) in which he suggested there was a phenomenon of Darwinian competition between the components of the organism. This theory remained largely unrecognised and Roux abandoned it to adopt a deterministic point of view instead. In the 20th century, Darwinism was applied to other applications in specialised areas of biology. In immunology, antibody synthesis is the result of a selective mechanism in which the variability of the genes that make the antibodies means that each immune cell synthesises a different antibody. The antigen only stimulates multiplication of the cell synthesising the antibody that neutralises the antigen itself (Jerne, 1955).

[31] The Fragments of Empedocles are published online at: http://philoctetes.free.fr/empedocle.html. See fragments 57–61.

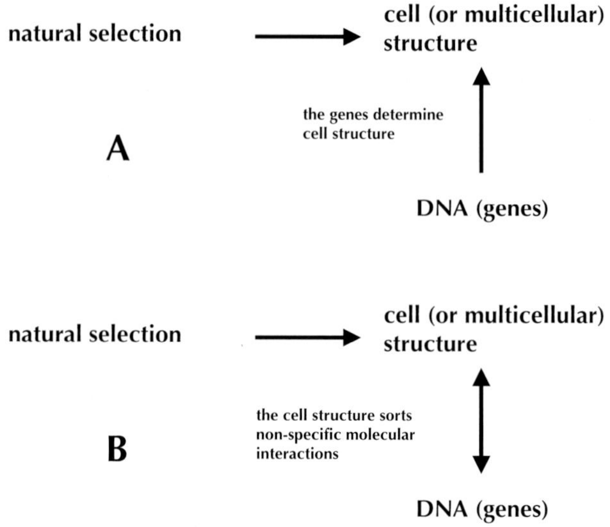

FIGURE 14. **Extension of evolutionary synthesis. A:** In the present context of evolutionary synthesis, natural selection is exerted on phenotypes (cells and multicellular organisms). It selects the most adapted variants and thus fashions the structure of the cells. In this way the genes (mutations) corresponding to the most adapted phenotypes are also selected. However, the relationship between the genes and the phenotype is unidirectionally determined. The genes determine the phenotypes. The processes of natural selection and ontogenesis are two separate processes. **B:** The cell structure, produced by natural selection, sorts the molecular interactions, which means that natural selection is a cause of ontogenesis. In this case, the relationship between the genes and the phenotype is no longer unidirectional but bidirectional. The genes provide the proteins while the phenotypic structure sorts their interactions from among the possible combinations. The two processes of natural selection and ontogenesis are but one single process of ontophylogenesis.

In the nervous system, the construction of neural cell circuits also seems to occur through 'neuronal selection'. In the first instance neurons are said to associate randomly owing to the huge number of combination possibilities for their ends (synapses and dendrites), creating very many circuits. Later, only the circuits that permit an adequate response to the stimuli received by the organism are retained (Changeux *et al.*, 1973; Edelman and Mountcastle, 1978).

However, despite these notable exceptions, embryogenesis and physiology have always been dominated by deterministic theories.

Our theory goes further. Not only do we suggest that the fundamental mechanism of ontogenesis is conceptually analogous to natural selection because it combines molecular chance and cell selection, but we also think that this mechanism is a true extension, within cell populations which make up the internal environment of living organisms (Bernard), of natural selection that produces the evolution of species (Darwin).

This is what we shall discuss in this chapter, in which we shall see how our theory resolves the contradiction in genetic determinism by uniting ontogenesis and phylogenesis. We will start from the more abstract principles and progress gradually towards the more specific mechanisms explaining cell differentiation and gene expression.

6.1 Ontogenesis and phylogenesis are but one process

The non-specificity of molecules has an inevitable consequence which must be taken into account in understanding ontogenesis: it introduces randomness into interactions between proteins. The great number of interactions possible from a set of molecules gives rise to numerous potential structures, not just a single one as in self-assembly or self-organisation. Each structure is an occurrence of a set of possibilities which each has only a certain probability of being produced. Consequently, the unique adult individual which results from embryogenesis is not produced by a simple mechanism of spontaneous assembly of molecules. Another mechanism must be applied to the potential combinations of interactions between molecules to restrict them and only select one of the structures possible, which will relate to the unique adult individual (Kupiec, 1983, 1996, 1999). Living beings are the result of such a process in which the very many interactions between non-specific molecules are subjected to natural selection. It can be illustrated by a thought experiment that is not intended to be realistic but explains this principle, diagrammatically.

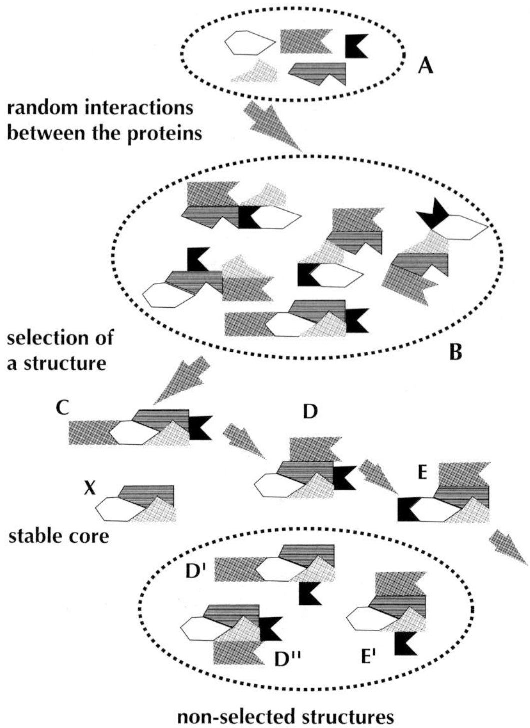

(devised from Kupiec, 1999)

FIGURE 15. Principle of ontophylogenesis. A: The five polygons represent proteins which are not stereospecific. **B:** They can associate with each other in different random ways and produce several structures. **C:** These structures do not have the same properties. In a given environment, one of them is selected because it is better adapted. **D and E:** This structure is never totally stabilised. It is partially modified owing to the random interactions of the proteins (D, D′, D″, E, E′), but its central core does not change because it is more stable (X). One of these potential structures may be selected depending on changes in the environment.

Let us imagine five proteins (Fig. 15A) capable of randomly combining and giving rise to several potential structures. These structures represent a cell in different states of differentiation (Fig. 15B). The stability and probability of each structure being formed depends on the probabilities of molecules encountering each other and the stability of their interactions. A single structure is selected from all the potential structures (Fig. 15C), which is the most stable (or the most viable) in the environment where the molecules are suspended. Because of the stochastic character of interactions between molecules, this ontogenesis never completely ends, but continues indefinitely since dissociation and reassociation is always taking place between the molecules, caused by thermal agitation. As long as the selective constraint (the environment) remains the same, the same structure continues to be selected. If the selective constraint were to change (if the environment changed), random interactions between the molecules would bring about modification and adaptation to this new environment. However, due to the pre-existing structure, the number of potential interactions i.e. the differentiation potential, offered by the random set of molecules is restricted. The structure is randomly modified, but not completely reconstructed. One of its parts, its central core, is more stable because it involves more interactions between the proteins, and therefore more bonds to maintain its cohesion (Fig. 15, X). On the other hand, random modifications are more frequent at the two ends of the structure because there are not so many bonds to stabilise it (Fig. 15D, D′, D″). The range of possibilities depends therefore on the state of the structure because the latter promotes certain interactions between molecules and prevents others. We suggest that in an actual cell its structure acts in a similar way. By sorting non-specific molecular interactions, this structure ensures its own maintenance and reproduction and also permits subsequent differentiation because random molecular interactions are not completely eliminated. In our imaginary example, one of the potential structures is selected (Fig. 15D) due to a new environment, and each time the environment changes again, the structure is modified within the framework permitted by its previous state and in line with the same mode of operation (Fig. 15E).

In this context, the contradiction in genetic determinism is resolved. The structure of the cell sorts random interactions between molecules without being the product of them. It is the result of a process that includes both stochastic molecular interaction and selective constraint. The macroscopic state of the cell at a given moment is not the product of its molecular state. Both depend on their previous state and on the randomness of the interactions between the molecules which make up the cell. Taking the history of the structure and the random character of molecular interactions into account in the causal explanation, to the detriment of a purely deterministic explanation based on levels of organisation and founded on stereospecificity, frees us from the contradiction of genetic determinism.

This outline is obviously an extreme simplification but it illustrates the general principle of ontophylogenesis: evolution over time of the structure is the continuation of its ontogenesis which is never completed. Ontogenesis and phylogenesis are but one process (Kupiec, 1986, 1997).

From this general principle we can detail three factors which have an influence on the ontophylogenesis of a living organism: DNA, the environment and the past.

1) DNA influences the probabilities of protein interaction

Proteins are subjected to thermal agitation and are moved by random Brownian motion. The probability of their encountering each other depends on their concentrations and their diffusion coefficients in the medium they are in inside cells. The more numerous they are and the more rapidly they diffuse, the more probable it is that they will encounter each other. Once achieved, an association between proteins is more or less stable depending on the strength of their bonds. These parameters are directly influenced in a cell by the DNA, its nucleotide sequence determining the amino acid sequence of the proteins, which itself influences their own binding and diffusion properties. In the same way, the concentration of proteins depends on the level of gene expression, which in turn depends on the structure of the genome.

2) The environment stabilises or selects certain cells rather than others

The strength of the bonds between molecules does not indeed depend solely on their intrinsic properties but also on the chemical composition and physical properties of the medium, such as its ionic concentration and temperature. In addition, each cell structure in a state of differentiation requires an optimal metabolic supply whether the supply is present or not in the environment.

3) Cells have a past on which they depend

A cell is the result at a given moment of its previous evolution. The molecular combination possibilities at the following instant are limited by its structure which promotes certain molecular interactions and prevents others, as seen in cellular compartmentalisation (see chapter 4, §4.3.2). There, the structure of the cell restricts the molecular combination possibilities and affects the future differentiation potential of the cell. When the structure changes through natural selection, under pressure from a change in the environment, its differentiation potential likewise changes.

Ontophylogenesis breaks with genetic determinism by including two factors absent from evolutionary synthesis — the stochastic character of interactions between molecules and the direct action of natural selection on ontogenesis. It does not however involve any emergence of irreducible properties. It is not a question of spontaneous self-organisation of the organism's components. Natural selection is a global constraint exerted on the organism, the origin of which is external to it.

6.1.1 *The model of the heap of cells and the origin of multicellularity*

How does the general principle of ontophylogenesis that we have just set out apply to multicellular organisms with differentiated cells? If it is the fundamental principle, it should shed light on how they function and the conditions in which they have appeared during evolution.

The simplest forms of multicellular organisms depend on the environmental and metabolic conditions in which the organisms live. For example, if placed in a medium poor in nutrients, the amoeba, *Dictyostelium discoideum*, will form a colony (an aggregate) of cells differentiated into two distinct types. Another form of multicellularity is even more primitive. The normally unicellular bacteria, such as *Escherichia coli,* also differentiate morphologically and genetically one from the other, when they grow in colonies in a poor medium. This differentiation depends on gene expression and the position of the bacteria within the colony (Ohgiwari *et al.*, 1992; Ben Jacob *et al.*, 1992; Shapiro, 1995).

These phenomena of multicellularity are simply explained by ontophylogenesis: cells differentiate to adapt to their microenvironment composed of other cells from the same organism. This mechanism of cellular adaptation is analogous to that of a living organism that adapts to its environment through natural selection. It is an extension of natural selection exerted within organisms (Kupiec, 1986).

To understand it we can think of it as follows: imagine a colony of cells growing on a solid medium containing its nutrient substrate (Fig. 16A). Its growth is subject to an obvious environmental constraint: the cells need to have access to the nutrient in the substrate to survive and multiply. Only the cells of the first layer in contact with the substrate have direct access to that nutrient. For the cells to be able to proliferate vertically, the nutrient has to diffuse vertically. The quantity that can diffuse is limited by the initial concentration in the substrate, so a decreasing gradient must be formed: the concentration of nutrient decreases vertically through the colony. This decrease is a consequence of the physical diffusion of the nutrient and of its consumption by the cells through which it passes.

From a certain height, the nutrient is no longer sufficient for the cells to be able to continue proliferating, and growth of the colony ceases (Fig. 16B). However, in metabolising the nutrient the cells produce metabolites. These metabolites likewise diffuse. In this heap of cells there are then other secondary gradients corresponding to the diffusion of the metabolites. Owing to these gradients, each cell finds itself located in a different microenvironment determined

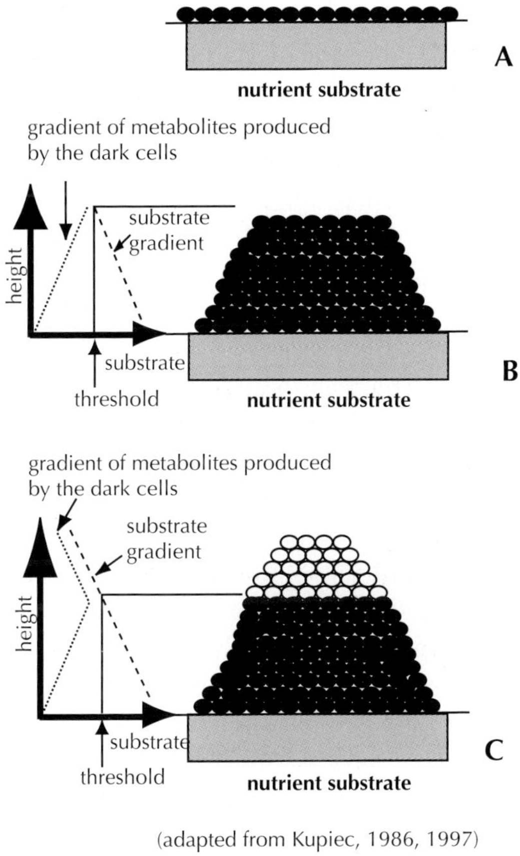

FIGURE 16. **Model of the heap of cells.** The growth of this cell colony depends on its environment. **A:** Only the cells of the first layer have direct access to the substrate. **B:** The other cells only have access to the substrate which is diffusing in the colony and forming a decreasing concentration gradient, or access to the metabolites, which form secondary gradients. Below a certain concentration threshold, the cells can no longer grow (dark cells). **C:** They must then differentiate to adapt to these microenvironmental conditions and continue to grow (light cells).

by the concentrations of nutrient and metabolites. The cells situated at the lower part of the colony have a microenvironment richer in nutrient while those at the upper part have a microenvironment richer in metabolites. To optimise their growth, the cells have to adapt to these microenvironmental conditions, which may occur randomly. Due to the non-specificity of molecules and thermal agitation, dissociation and reassociation takes place continually in the cells between different molecules. These events occur with certain frequencies determined by the structure of the cells, the concentrations of molecules and their speeds of diffusion (see earlier section in §6.1). If one of the cells situated at the upper part of the colony where growth has ceased (Fig. 16B) is subject to such stochastic remodelling in its chromatin and this remodelling induces the expression of new genes allowing it to use the resources which are accessible to it (i.e. the metabolites and a minimal quantity of nutrients), the growth of the colony can resume (light cells, Fig. 16C; for an example of stochastic remodelling see Fig. 21). This 'proto-organism' is then made up of two types of cell, corresponding to two 'tissues' with different 'functions'. One type, made up of dark cells, metabolises the nutrient and provides nourishment to the tissue made of light cells, by supplying it with metabolites. Metabolic cooperation is established between the cells which differentiate as a function of their position in the colony.

Every time our proto-organism develops in an identical environment the same structure is produced because it is subject the same selective constraint. However, if that constraint changes, that is to say, if the concentration of nutrient in the substrate changes, the concentration gradients within the colony will likewise change. The cells then have to adapt to these new conditions. The dark cells will grow wherever there is an adequate quantity of nutrient. The size of the two 'tissues' (dark and light) will therefore change because the quantities of nutrient and metabolites available in the colony will vary owing to modification of the gradients and, in the end, the very structure of the heap of cells will be modified to comply with these new gradients (Fig. 17).

This phenomenon of colony adaptation complies with the general principle of ontophylogenesis in Fig. 15, except that it is produced in the internal environment of the multicellular colony and concerns

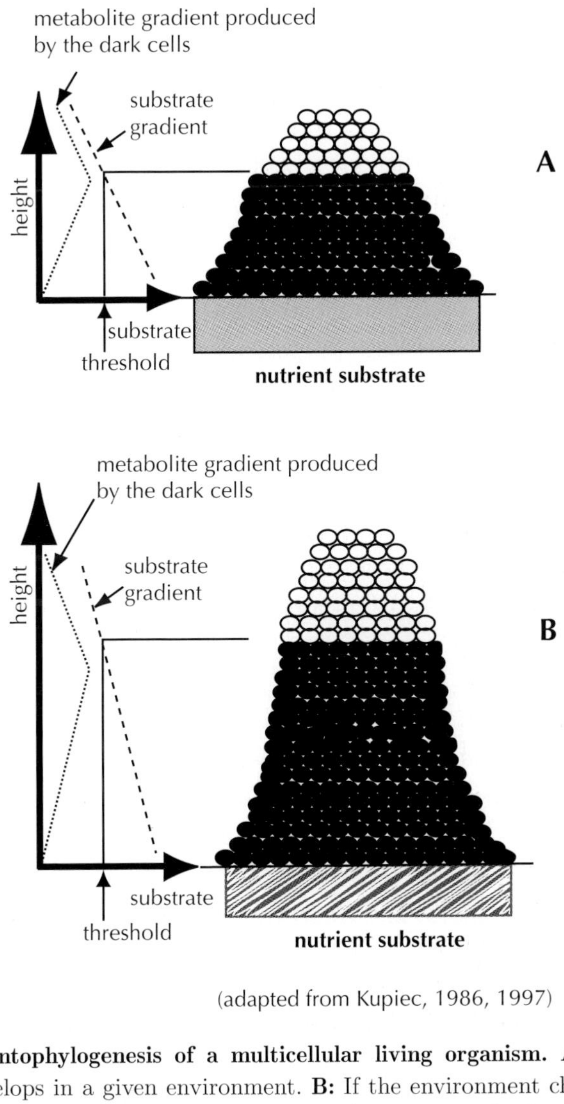

FIGURE 17. Ontophylogenesis of a multicellular living organism. A: The proto-organism develops in a given environment. **B:** If the environment changes (in this example, if the substrate concentration changes), the internal selective constraints (the substrate and metabolite gradients) which depend on it likewise change. The threshold necessary for growth of the dark cells moves. This results in modification of the structure of the cell colony. One single mechanism explains its development and its evolution.

populations of cells. Selective constraint is propagated in the proto-organism and determines the microenvironments to which the cells stochastically adapt. By using the same mechanism as that producing ontogenesis, the proto-organism evolves to adapt when the environmental conditions change. Its ontogenesis and phylogenesis are produced at the same time by a single process of ontophylogenesis.

We took the example of a colony growing on a solid substrate, but the same metabolic constraint applies for a spherical colony growing in a liquid medium. In this case, diffusion occurs from the exterior towards the centre of the colony, and the gradients follow this axis. The concentration of nutrients decreases as they pass from the exterior towards the centre while conversely, the metabolite concentration increases. The region poor in nutrient where cell growth ceases is therefore the centre of the sphere. The embryos of many organisms in the first stages of embryogenesis form a ball of cells like this and precisely in their centre a cavity forms, called the blastocoel.

The general idea that cells differentiate depending on their position in the organism, as in our example of the heap of cells, is not new. It is the basis of Lewis Wolpert's theory of positional information and has been expressed by other authors in various forms (Wolpert, 1969, 1989, 1991). As regards the role of metabolism, Charles Manning Child (1869–1954) undertook considerable experimental and theoretical work to demonstrate that the existence of metabolic gradients in the embryo is the causal factor of its development (Child, 1941). The role of ontogenetic or phylogenetic constraints is likewise well-known (Maynard Smith *et al.*, 1985; Arthur, 1988; Williams, 1992). All these theories are nevertheless deterministic. Ontophylogenesis is distinct from them by virtue of its probabilistic, selective and unified conceptual context. The question of the level at which selection operates (whether at the molecule, the cell or the organism) also recurs in evolutionary biology (Dawkins, 1976; Brandon and Burian, 1984; Williams, 1992). In this respect, Leo Buss (1987) also suggested that natural selection is applied to cells and that multicellularity arises from competition between cell lineages. His analysis demonstrated in a very detailed way how this cell selection explains the transition between unicellular and multicellular organisms. However, it does not break with genetic determinism, as does ontophylogenesis.

The model of the heap of cells is very basic but illustrates a general principle which explains the origin of multicellularity and the context in which it develops. It is evident however that from this origin the structure of organisms has become more and more complex over the billion years that they have been evolving, and that this increase in complexity has optimised their functions. Consequently, in multicellular organisms with a blood circulation system distributing nutrients to organs situated at some distance, the elementary logic that we have described certainly does not suffice, but the increase in the complexity of organisms during evolution does not nevertheless invalidate ontophylogenesis. The appearance of a circulation system only emphasises the importance for multicellular organisms of being able to distribute resources to all the parts of the organism. It in no way implies that older mechanisms have been eliminated.

Evolution does not work like an engineer who rationally replaces the parts of a machine. It often leads rather to traits or mechanisms that have appeared at different periods of the evolution of the organism being superimposed, rather than their being purely and simply replaced. At the early stages of embryogenesis, even in a complex organism, nourishing the tissues has to continue in line with a model like the heap of cells, all the while the embryo is a collection of cells in the process of differentiation with the organs not yet in place. At the adult stage, the model of the heap of cells is still relevant to explain tissue organisation within certain organs irrigated by the blood system. The mammalian liver provides such an example. It is divided into three areas each with a different function corresponding to gradients of nutrients arising from afferent blood vessels which supply blood rich in oxygen, nutrients and hormones. Across these three areas, there are also gradients of enzyme activity which correlate with the nutrient gradients and enable the nutrients to be used optimally. Genetically, these gradients of enzyme activity depend on differential gene expression (Jungermann and Sasse, 1978; McGrane *et al.*, 1992).

During evolution, organisms have not only undergone an increase in the complexity of their structure, but also optimisation of their regulation. Molecules such as hormones that act like signals have arisen. Later in this chapter we will discuss in detail cell differentiation and gene expression mechanisms which include the action of these signals in the context of ontophylogenesis. Before that, we must look in depth at two theoretical points which shed light on the process of increasing the complexity of organisms and show that ontophylogenesis is not only consistent with Bernard's classic theory of the internal environment but in many respects is a development of it.

6.1.2 *The organism interiorises its environment*

Ontophylogenesis is a phenomenon of hetero-organisation. In the model of the heap of cells, the environment is propagated *via* the nutrient gradients inside the organism forming cell microenvironments. The structure of the organism arises owing to the environment being interiorised within it in this way, and is inseparable from it.

This concept runs counter to the notion of an intrinsically autonomous individual as conveyed by genetics and self-organisation, theories which hold that the relationship of the organism to its external environment is limited to the supply of nutrients that allow it to form as individually determined within itself (by its genes or the emergent properties). Natural selection indeed acts on the organism, but in the adult stage, once it is already formed, and not on its embryogenesis as is the case in ontophylogenesis. It would thus be in the nature of embryogenesis to create an organism separated from the environment by a watertight barrier, ensuring its existence thanks to its internal environment being constant. The separation, in this finalist conception, of the living organism from the environment is an essential characteristic of life. In actual fact, this is a misinterpretation which distorts Bernard's theory of the internal environment, which has been forgotten, but deserves to be reconsidered.

For Bernard, life is not an intrinsic characteristic of living matter but a phenomenon which only exists in its relationship to the environment. "*In the same way, life results from contact of the organism with its environment; we can no more understand it through the organism alone than through the environment alone*" (ISEM p. 75).[32] In this relationship the organism is bound to adapt to its external environment. Indeed, "*It is not by warfare against cosmic conditions that the organism develops and maintains itself, but on the contrary, by an adaptation, an accord with these*" (LPL, p. 48). It is in the context of this adaptation that the internal environment plays a fundamental physiological role. It serves as an interface for transmitting the influences of the environment on the organs, for "*Only by passing into the inner, can the influence of the outer environment reach us...*" (ISEM, p. 76). Bernard sees three distinct modes of adaptation indicating increasing degrees of autonomy of the organism and through which the internal environment becomes progressively more constant. 'Latent life' corresponds to a complete lack of autonomy. Some living organisms are totally dependent on external conditions, and if they are not appropriate these organisms enter a state of latency which they leave only when the external conditions become favourable again (LPL pp. 48–77). This is the case of organisms which sporulate or form seeds that no longer exhibit the least biological activity. 'Oscillating life' corresponds to that of living organisms whose mode of existence can vary greatly depending on environmental conditions, without ever reaching the state of latency (LPL pp. 77–83). This is the case of cold-blooded animals that hibernate, in whom life slows down in winter and becomes active again in the spring. The third form of adaptation is 'constant life' (LPL pp. 83–91), and corresponds to living organisms that have a constant internal environment and are thus autonomous in their relationship to the environment. However, it is

[32] This is reaffirmed several times, particularly in the "Lectures on the Phenomena of Life". See the second lecture devoted to the internal environment.

totally incorrect to interpret the constitution of a constant internal environment as a finalist or teleonomic process caused by a property intrinsic to living beings endowed with the prior intention of creating individuals.

> *"In constant life, the living organism seems free, and vital manifestations appear to be produced and directed by an inner vital principle free from external physicochemical conditions; this apearance is an* ***illusion.***[33] *On the contrary, it is particularly in the mechanism of constant or free life that these close relations*[34] *exhibit themselves in their full clarity"* (LPL, p. 91).

Bernard speaks of illusion because the constancy of the internal environment is a mechanism of adaptation of the organism to the environment, in relation to which and precisely because of it, it remains dependent. The living organism only exists in relation to the environment and the constitution of the internal environment is a consequence of this relationship, which requires mechanisms at the interface between the organism and its environment permanently compensating for variations in the latter. This indicates an increase in the complexity of the relationship to the environment and not absolute independence expressing an internal teleonomic property, inherent in living organisms.

In addition, once the internal environment has been formed during embryogenesis, the organs lead an autonomous life there. This very radical aspect of Bernardism has been completely overshadowed by modern biology, which considers that the organism is governed centrally by the genetic programme. Bernard declares, on the contrary, that its parts are autonomous within the internal environment.[35] This is a very important point which he reasserts

---

[33] Original text not in bold.

[34] With the environment. (Not in original text.)

[35] This principle is based on experimental observations likewise described by Bernard. For example, organs are anaesthetised by the cold directly by the effect of the temperature: the anaesthesia is not controlled by the central nervous system.

several times in 'Lectures on the Phenomena of Life' since it is because of this autonomy that the parts of the organism act on each other and can organise themselves so as to function.

> *"Thus differentiated and specialised, the anatomical elements live their private lives in the place they are assigned, each according to its nature. (...) It is by the mediation of the interstitial fluids, forming what I have called the 'milieu intérieur', that the solidarity of the elementary particles is established, and that each one receives the repercussions of the phenomena that take place in the others. The neighboring elements create for the one under consideration a certain ambient atmosphere, and it feels the changes in it, which regulate its life"* (LPL p. 260).

In ontophylogenesis, cells are autonomous and adapt to the internal environment in which they are situated, which is itself dependent on the external environment. From this point of view, therefore, the two theories concur perfectly. Ontophylogenesis however also explains how organisms become more complex as they pass through the three stages described by Bernard.

An organism always depends on the DNA, the environment and the past borne by the structure, although the relative influence of each of these factors will have been modified during evolution. The initial cell which gave birth to the ontophylogenesis process will have been subjected, during its history, to an endless accumulation of selective constraints which have progressively adapted its structure to more and more diverse environments, a process making it extremely complex and robust. The possibilities offered to the random set of molecules will have been reduced, owing to restriction on the number of potential molecular combinations caused by the increasing importance of structuring the cells, to the point where today's organisms are identically reproduced with a limited number of possible variations that are nevertheless sufficient to allow the cell differentiation necessary for forming tissues. Bertrand Laforge suggested a very simple diagram, from which we will draw inspiration here, to illustrate how constraints may force a mechanism to behave in an apparently deterministic way even though it is intrinsically probabilistic (Laforge, 2004).

Take a marble moving around completely randomly. In the absence of constraints (Fig. 18A), it describes a movement which is totally unpredictable in advance. It can explore all the space around it. On the other hand, if it is enclosed within two walls, this constraint restricts its possibilities for moving around because it hits the walls and stays within the space delimited by them. Its movement therefore becomes partly predictable. We can foresee the area in which it will move (Fig. 18B). In an extreme case, if the constraint created by the two walls increases, i.e. if they are very close to each other, the marble will move in a straight line though its movement is still intrinsically random (Fig. 18C). In this case, the overall movement of the marble is totally predictable (it will always describe a straight line), although it will go forwards and backwards randomly between the two walls. If this experiment is repeated, the result will always be the same: the marble will trace a straight line. From one experiment to another however, there will be stochastic

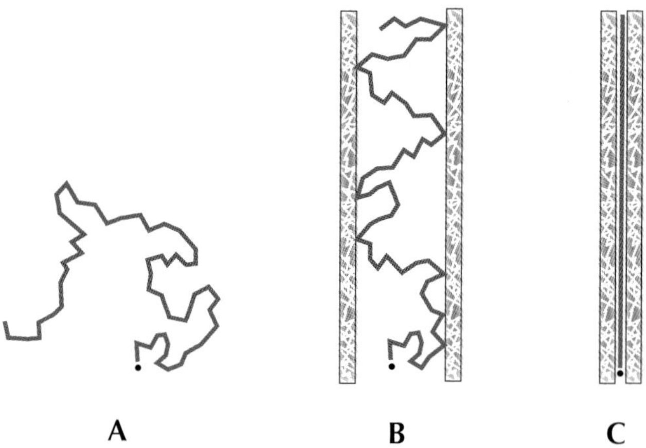

FIGURE 18. The effect of constraints on a random phenomenon. A: A marble moves randomly. **B:** The constraint of the walls limits its possibilities for randomly moving around in a known space. **C:** In the extreme case, the marble always describes a straight line although its movement is intrinsically random.

variations in the way it does this. Cell and functional structures act in the same way in an organism to reduce the macroscopic effect of the random behaviour of the molecules, without nevertheless eliminating it. Proteins are always subjected to thermal agitation, which permits stochastic variations creating tiny differences between cells that are the origin of their differentiation.

In the course of evolution, under the pressure of natural selection in which environmental constraints with an effect on organisms accumulate, cell structuring has greatly increased. A eukaryotic cell of a multicellular organism has many more organelles and membranes than a unicellular prokaryotic cell, and this structuring has reduced molecular randomness. Multicellular organisms have also become more complex through the creation of structures and functions. Functional structures ensuring the constancy of the internal environment have been selected because they increase the viability of the organisms. At the same time, the roles of the DNA and the structure have come to predominate and the organism has thus escaped from variability of the environment.[36]

There are, of course, notable divergences between ontophylogenesis and Bernard's theory. He did not apply his theory to the problem of ontogenesis which he considered to be too complicated, and he was a staunch adherent to determinism (Gayon, in press). At the same period variability was already present, on the other hand, in Darwin's theory. Darwin thought that it was the variations in organisms that made it possible for them to adapt to their environment. Ontophylogenesis therefore combines aspects of Darwin's natural selection with Bernard's internal environment.

6.1.3 *The organism functions for the cells, not the reverse*

The relationship between the organism and its parts is another basic question which often leads to confusion. In ontophylogenesis,

[36] Although its influence remains of considerable importance for many species, particularly plants.

cells are involved in a contradiction: on the one hand they are individualists, each one optimising its own multiplication, while on the other, they each need the others as each uses the products of metabolism of its neighbours. Each cell, although working for its own good, is subordinate to the whole. It does not enjoy total freedom as its freedom is limited in that the cell is constrained to differentiate in a way appropriate to the place it occupies in the society of cells. The organism is constructed in the light of this contradiction which only reflects its own relationship to the environment. It exists as an individual because there is functional unity implicating all the relationships involving exchange between the cells. At the same time, there is no final aim in the organisation established of creating the organism for its own sake as an individual unit. It is the consequence of a process which ensures as best it can the life of cells. This is illustrated by the example of the heap of individual cells. Those in the dark layer nourish the cells in the light layer but they do not perform this function in any finalised way. On the contrary, they are only functioning for their own good. The function they acquire is a consequence of their metabolic activity and the internal relationships which become established in the cell population. These relationships themselves only exist because of the relationship to the environment which polarises the exchange of nutrients.

Such a conception again challenges our firmly held beliefs. Since we are spontaneously anthropocentric, we have a strong tendency to believe that we are the finality of natural processes, whether in evolution or embryogenesis. As regards evolution, the theory of natural selection puts an end to this belief despite all the religious resistance which that theory incited. As regards embryogenesis, this belief is perpetuated with the theory of genetic programming where the final cause has become known as teleonomy (Pittendrigh, 1958; Mayr, 1961; Hull, 1982). The aim of the genetic programme is to construct the organism according to the instructions present in the genome, all its parts, from the molecule to the organs, being dedicated to this project (Fig. 11A).

We find it difficult to accept the idea that we are at the service of our cells, rather than the reverse. Nevertheless, as with the previous notion, this idea is to be found in the theory of the internal environment. Bernard's developments on this subject can even be applied here word for word. The following quotations are somewhat long but they are important, given the confusion that reigns in debates nowadays in biology.

> *"We have said that the living organism is an association of cells or elements more or less modified and grouped into tissues, organs, apparatus and systems. It is thus a vast mechanism resulting from the assemblage of secondary mechanisms. From the monocellular being to man, all degrees of complication are encountered in these groupings; organs are added to organs, and the most highly developed animal possesses a great number of them that form the circulatory system, the respiratory system, the nervous system etc.*
>
> *It has been believed for a long time that these superadded mechanisms had their own* raison d'être *or that they were the result of the caprice of an artistic nature. Today we ought to see in them a growing complexity regulated by law. Anatomy, restricting itself to the observation of forms, did not succeed in deriving it. It is physiology alone that can give an account of it.*
>
> ***Organs and systems do not exist for themselves; they exist for the cells, for the innumerable anatomical elements that form the organic edifice.***[37] *The vessels, the nerves, the respiratory organs appear as the histological framework becomes complicated, so as to create around each element the environment and the conditions that are necessary for this element, so as to dispense to it in appropriate measure the materials that it needs; water, food, air and, heat. In the living body these organs are like the factories or the industrial establishments in an advanced society which provide the various members of this society with the means of clothing, heating, feeding, and lighting themselves.*
>
> ***Thus the law of the structure of organisms and of organic development is bound up with the law of cellular life. It is to make possible and to regulate more closely the life***

[37] Original text not in bold.

> ***of the cells that organ is added to organ, and apparatus to systems. The task imposed upon them is to bring together in quality and quantity the conditions for the life of the cells"***[38] (LPL, p. 259).

Organisation does not aim to ensure the organism functions as a whole. It simply creates the internal environment which provides the cells with what they need to survive.

> *"In résumé, life resides within each cell, in each organic element, which functions for its own account. It is not centralised in any part, or in any organ or apparatus of the body. All these apparatus are themselves created in the light of the life of the cell"* (LPL, p. 265).

Cells are however at the same time subordinate to the organism through the very constitution of this internal environment. "*It is by the infinite variety that the internal environment presents from one place to another and by its particular and constant composition at a given point that the subordination of the parts to the whole is established*" (LPL p. 261). This is a very important aspect. "*It is the subordination of the parts to the whole that makes an integrated system, a whole, or an individual, out of the complex being. It is in this way that unity is established within living bodies*" (LPL pp. 262–263). This subordination of cells is demonstrated in tissue graft experiments which only succeed if they are done at the right place in the organism, and by regeneration experiments in which the animal rebuilds its original form after removal of a part (LPL pp. 260–266). The organism indeed exists, therefore, but it is the result of cell life, not its finality. Seen in this light, ontophylogenesis helps us to understand how during the course of evolution we have been able to go from selection acting on individual cells to selection acting on the whole organism. Let us imagine that in the heap of cells (Fig. 16), the structure of the dark cells is randomly slightly modified in such a way that it does not alter their capacity to metabolise the nutrient

[38] Original text not in bold.

but facilitates the transfer of its metabolites to the light cells. In real organisms, a similar function is fulfilled by the cells of the heart distributing blood throughout the body of an animal. Such a modification would not directly benefit the dark cells as individual cells, but would, on the other hand, promote the growth of the light cells and consequently that of the heap of cells as a whole. The dark cells would thus benefit indirectly. Their growth would be optimised as cells belonging to an organised set. The variation in the structure of the cell giving rise to this transformation would be selected because it is favourable to the whole of the population.

This model provides us therefore with a general framework for understanding, without finalism, the appearance of functions which improve the performances of the organism as a total individual entity. It provides an explanation for Bernard's belief in which he asserted that "*The actual role of organs is not the agent that has caused their formation*" (LPL p. 243). This antifinalism of Bernard's, to which we wholly subscribe, goes even further since it results in depriving the notion of function of any objective value, and in its complete 'de-essentialisation'.

> *"Apart from the intervention of the mind, and insofar as there is objective reality, there is in the organism only a multitude of acts, of material phenomena, simultaneous or successive, dispersed among all the elements. It is the mind that grasps or establishes their interconnections and their relationships, that is to say, their function. Function is thus something abstract which is not represented materially in any of the properties of the elements"* (LPL, pp. 268–269).

6.2 The deterministic theory of cell differentiation

We can now return in more detail to the question of cell differentiation, which we looked at earlier but at a very general level. Although embryogenesis cannot be reduced exclusively to this process, it is one of its main aspects. The problem it poses is in understanding how differentiated cells are produced from a single cell (the germinal cell) and how they are organised into ordered cell tissues. The deterministic conception which predominates at the present time is influenced by genetics and by the early work on experimental embryology.

6.2.1 *Embryonic induction*

Driesch's experiments demonstrate that the differentiation of cells cannot be solely the expression of their internal determinants (see chapter 5, §5.3.1). An additional mechanism restricting their development potential is therefore necessary and even if we reject Driesch's and Elsasser's vitalism, we must explain it. Light has been shed on this question by the work of Hans Spemann (1869–1941), who performed tissue grafts inside embryos which demonstrated the primordial role of interactions between cells. In the course of the embryo's development, the cells have a mutual effect on each other. What each cell becomes depends on the influences it receives from the other cells, a phenomenon which has been called 'embryonic induction' (Spemann, 1938; Bouwmeester, 2001).

Although Spemann's experiments did not in themselves indicate the nature of the induction mechanism, it was immediately conceived as a deterministic phenomenon. It was supposed that embryonic cells produce induction molecules which act on their neighbours determining how they will differentiate (Saha, 1991).

6.2.2 *The instructive model*

The deterministic conception of embryonic induction gave rise to what became called the 'instructive' model of cell differentiation. In this model, cells differentiate because they receive 'instructions' corresponding to signals (or information) carried by proteins. These signals trigger cascades of reactions inside the cells (see Fig. 8) which end in genes being activated producing cell differentiation.

In Fig. 19 cell B differentiates into cell D because it receives a protein signal *d* synthesised by cell A. In the same way, cell A differentiates into cell C because it receives a signal *c* synthesised by B.

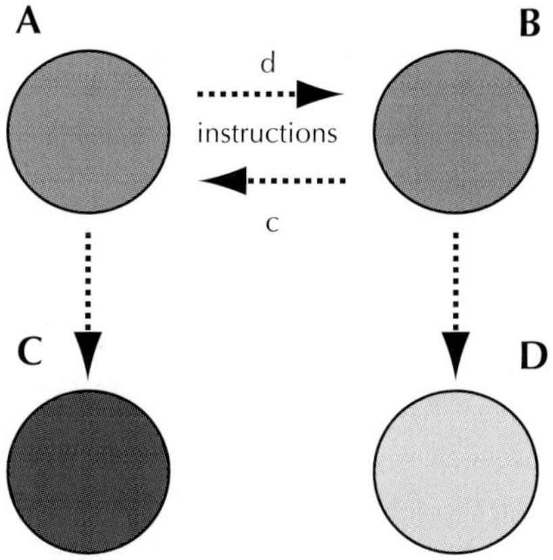

(adapted from Kupiec, 1983, 1997)

Figure 19. The instructive model of cell differentiation. Cells A and B differentiate specifically into C and D in response to the signal they receive (*c* and *d*). Embryonic development is supposed to be a succession of elementary stages identical to this, with each stage corresponding to the expression of specific genes encoding differentiation signals.

In this deterministic context, the differentiation of tissues during embryogenesis is seen as a series of elementary stages similar to that described in this figure, and because each stage involves the expression of genes encoding for instructive proteins, the entire process expresses the genetic programme (see chapter 3).

There are several variants of the instructive model. If the signals are carried by membrane molecules, there needs to be direct contact between the cells at their membranes. If the signals are carried by diffusible molecules, they can act from a distance. In certain cases, instructive molecules may also form concentration gradients and exert their specific effect only at a specific concentration. The signal may equally correspond to a combination of molecules, not one only, but whatever the variant of the model, its principle remains the same: cells differentiate in a way determined by the signals they receive.

There are decided advantages to this model. Firstly, it integrates the very large amount of data which reveal interactions between cells, the transduction of signals and the differential expression of genes. Secondly, a deterministic model might seem perfectly relevant at first sight for accounting for the precision in the way an organism functions. However, if we analyse it in more detail, it reveals difficulties which must lead us to challenge it.

6.2.3 *The instructive model trips up against the contradiction in genetic determinism*

By definition, the signals must be specific since this is a deterministic model. In Fig. 19, cell B reacts in a unique way to signal *d*, and it is the same for cell A in relation to signal *c*. This necessarily presupposes that these signals also induce specific reactions within cells. If this condition is not fulfilled, the model can no longer function because several cell reactions are possible for a single signal. Now, despite intense research efforts undertaken since Spemann's work was published, these specific signals have never been discovered. Of course, very many molecules implicated in embryogenesis have been isolated, but as we saw earlier, they do not exhibit this character of specificity predicted by the model (see chapter 4). To overcome this difficulty, adherents to the deterministic model usually argue that it is not just one molecule that transmits the specific signal, but a combination of cofactors. Considering again the example in Fig. 19, this means that the signals *d* and *c* no longer correspond to single molecules synthesised by cells A and B, but to sets of molecules synthesised by these cells. In reality, this is an unsatisfactory palliative explanation for it only moves the problem elsewhere. Indeed, if cells A and B already synthesise sets of different molecules, it means that they are already differentiated whereas this is precisely what the model is supposed to be explaining. Once again we stumble against the contradiction in genetic determinism which consists in reversing the cause and effect. This contradiction is moreover inherent in the model itself which is based on initial asymmetry between cells which synthesise different signals. The model must therefore presume a diversity of cells, the appearance of which it is supposed to explain. The usual practice for resolving this contradiction is to evoke the effect of morphogenetic gradients pre-existing in the egg. Owing to their heterogeneous distribution, there is said to be unequal distribution in each daughter cell of the molecules present in the egg after each of its divisions. This mechanism is supposed to create initial differentiation of the cells which would set differentiation in motion according to the instructive model. Although morphogenetic gradients are an indisputable reality and play a certain role in embryogenesis, this explanation is still a palliative measure which does not really resolve the contradiction. It bases the whole of embryogenesis on the egg's initial gradients. A great many experiments, however, including those performed by Driesch (see chapter 5), demonstrate that embryo cells that have undergone several successive divisions are still capable of reconstituting complete organisms, even though the morphogenetic gradients have been destroyed in these cells. Regulatory mechanisms must therefore exist that are capable of creating heterogeneity of cell types in the course of embryogenesis.

6.2.4 *The instructive model does not account for variability in cell differentiation*

For a long time cell differentiation for a great many cell lines has been analysed using a variety of techniques. These analyses have revealed stochastic variability in the differentiation of numerous tissues which is not compatible with a deterministic model. This variability may be manifested in various ways: either each individual cell of a single population differentiates with a chronology that varies from one cell to another, or the descendants of the individual cells vary as regards their differentiated cell content. Such variability, which can only be modelled using stochastic models, has been observed in a variety of organisms, *ex vivo* and *in vivo*, with a number of experimental techniques and for many cell lines. Jim Till and his team (1964) put forward the first stochastic model to account for variability in the differentiation of haematopoietic stem cells. Following this, similar work has been performed on cells of the immune system, blood, skin, liver, bone, intestinal and heart cells, as well as on embryo cells at the blastula and neural crest stages (Till *et al.*, 1964; Nakahata *et al.*, 1982; Godsave and Slack, 1991; Barrofio and Blot, 1992; Yamada *et al.*, 2007; Bennett, 1983; Lin *et al.*, 1994; van Roon *et al.*, 1989; Bohme *et al.*, 1995; Davis *et al.*, 1993; Paulus *et al.*, 1993). This list is not exhaustive.

6.2.5 *The instructive model does not account for stochastic gene expression*

A great many molecular processes could be the origin of variability in cell differentiation. Thanks to techniques which can analyse gene expression in individual cells, a vast amount of experimental data has now accumulated which demonstrate that this phenomenon is itself a stochastic process and that it is here that variations giving rise to cell differentiation could be produced.

According to the deterministic model, identical cells in a uniform environment ought to express the same genes, but that is just not the case. There are always differences of expression between individual cells. A gene expressed in one cell is not necessarily expressed in another cell of the same population. This was originally demonstrated with regard to the expression of the proteins of many cellular or viral genes (Ko *et al.*, 1990; Ross *et al.*, 1994; Fiering *et al.*, 1990; White *et al.*, 1995; Takasuka, 1998). In muscle cells with several nuclei, the different nuclei, while nevertheless sharing a single cytoplasm, do not express the same genes. It has been demonstrated in this case that the variability of expression is located directly at the level of the gene transcription into RNA (Newlands *et al.*, 1998). We have also detected variability in the transcription into RNA of the insulin receptor gene (Heams and Kupiec, 2003).

These results have been confirmed by even more spectacular experiments revealing differences in expression between the two chromosomes of a diploid cell which each carry an allele of the same gene. Such heteroallelic difference in expression has been observed for many genes (Chess *et al.*, 1994; Wijgerde *et al.*, 1995; Held *et al.*, 1999; Holländer, 1999; Rivière *et al.*, 1998; Jouvenot *et al.*, 1999;

Nemazee, 2000) and it is difficult to see how it could be compatible with the deterministic model. The latter indeed predicts that genes in the same microenvironment, and therefore influenced by the same transcriptional regulatory factors, should be in the same state of activity or inactivity. Yet this is not the case for all these genes. Their alleles are located in the same nucleus but they are not expressed in the same way on the two chromosomes that carry them. The simplest interpretation to explain this is that gene expression is a stochastic phenomenon. An allele is expressed on one chromosome but not necessarily on the other at the same time, for each allele only has a certain probability of being expressed at a given moment. This explanation has been confirmed by experiments, in which in the bacterium *Escherichia coli* and the yeast *Saccharomyces cerevisae*, two copies of a single gene, artificially placed in strictly identical intracellular contexts, are not transcribed in the same way at a given moment. Their expression undergoes major stochastic fluctuations (Elowitz *et al.*, 2002; Raser and O'Shea, 2004).

All these data suggest a probabilistic interpretation, advanced by most of the authors. Activation of gene transcription would seem to be limited by assembly of the protein complex which initiates it. Very many different molecules participate in forming this complex but there are very few of some of these proteins in the cell nucleus. As they have to diffuse right to the transcription initiation site to assemble, this is a rare occurrence and there is only a small probability of it occurring at any given moment. When it does occur, after the gene has been transcribed and the complex has dissociated, it is repeated after a period which varies randomly from one time to the next (McAdams and Arkin, 1999; Hume, 2000; Paldi, 2003; Coulon *et al.*, in press). A considerable amount of work has been done to analyse this phenomenon precisely and quantitatively. The result from it is that the stochastic expression of genes is nowadays considered an unquestionable fact (Kaern *et al.*, 2005; Raser and O'Shea, 2005; Kaufmann and van Oudenaarden, 2007; Heams, in press). From a philosophical point of view, it can be interpreted objectively (Merlin, in press).

Now, it has to be said that this raises a critical question. Gene expression is the elementary stage of the genetic programme: how is it possible for it to be a stochastic phenomenon while the genetic programme is deterministic by definition? Two answers to this can be envisaged. Either stochasticity is noise affecting the controlled functioning of the genetic programme, in which case the cell must have developed mechanisms to eliminate its potentially negative effects, or normal gene expression is really an intrinsic stochastic process and the cell exploits this stochasticity to bring about its functions. In this case the idea of a deterministic genetic programme would have to be abandoned. The latter option seems to have been reinforced by a series of experiments, which have indeed shown that it is the stochastic expression of genes that is at the origin of cell differentiation in unicellular and multicellular organisms (Becksei *et al.*, 2001; Blake *et al.*, 2003; Isaacs *et al.*, 2003; Wernet *et al.*, 2006; Maamar *et al.*, 2007). However, these results require to be further generalised, above all to embryonic development.

6.3 The Darwinian theory of cell differentiation

6.3.1 *From differentiation to cell identification*

In the instructive model, cells differentiate due to the influence of signals they receive. The Darwinian model is based on quite different logic. It takes as its starting point the non-specificity of proteins which induces a number of stochastic molecular interactions, the occurrence of any of which is probable.

In an individual cell, certain of all the interactions that are possible take place, but owing to their probabilistic character, the same interactions do not occur in each cell of a cell population. Cell diversity originates from this probabilistic process.

In the diagram shown in Fig. 20, the two cells A and A′ are identical. Two random molecular events may be produced there, *b* or *c* (each corresponding to non-specific interactions between at least two molecules. For an example, see Fig. 21). These events permit the stochastic expression of two different sets of genes and

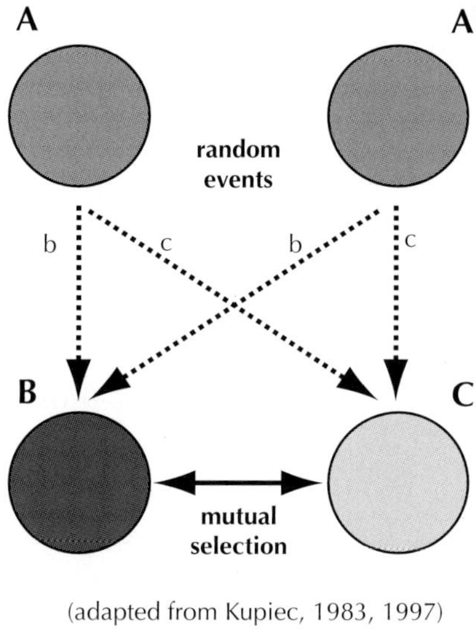

FIGURE 20. The Darwinian model of cell differentiation. The cells A and A′ differentiate into B or C depending on the random molecular interactions *b* or *c* which are produced. Cellular interactions are also involved in the selection or stabilisation of these randomly obtained phenotypes.

depending on which set is expressed, the cells differentiate into type B or C.

Interactions between cells do occur in this model, but their function is not to induce their differentiation as is the case in the instructive model. They select or stabilise the cells which differentiate randomly. In Fig. 20, the interaction between cells B and C leads to selection (or stabilisation) of B by C and of C by B, and therefore to their coordinated differentiation.

Unlike the instructive model, in the Darwinian model the precision of embryonic development is not based on the precision of molecular events, as prescribed by the principle of order from order. Two factors play complementary roles. Firstly, there is a population effect. While random events cannot be reproduced in the cell or the individual molecule, they are reproduced, statistically, in cell or molecular populations, in which they do so with stable mean frequencies subject to variations depending on the size of the population. The larger the size of the population, the less the variability (see chapter 2). Secondly, selection also refines precision because it imposes organisation between the types of cells. Thus in the model of the heap of cells (Fig. 16), multiplication of the dark and light cells is interconnected through indispensable metabolic exchange. In the same way, in Fig. 20, cells A and B are interdependent; one cannot exist in a stable manner without the other. Jim Till (1981) suggested a Darwinian model for the differentiation of haematopoietic cells. We also suggested this, from a theoretical analysis, as a general model of cell differentiation based on the stochastic expression of genes, first of all during a colloquium (Kupiec, 1981), then in an article (Kupiec, 1983). At that time, suggesting that cell differentiation and gene expression might be probabilistic phenomena was considered by most biologists as extremely eccentric, to put it mildly! Let us however concentrate on our discussion.

This model does resolve the difficulties of the deterministic model. It is based on the non-specificity of molecules which induces random events, it explains why a diversity of cells is produced from a homogeneous population and it integrates stochastic gene expression.

However, it in turn raises a question because it inverts the problem of differentiation, as it is usually posed. The Darwinian model implies that, owing to random interactions between molecules which are inherent in them, cells can change state and initiate differentiation without the intervention of an inducer signal. This prediction is compatible with experimental data which has shown that cells are spontaneously transformed in the absence of a signal triggering transformation when they are cultured outside the internal environment of the organism (Rubin, 1990). This phenomenon is well known to those cultivating cells and obliges them to clone cells regularly so that they keep their original characteristics; otherwise, they transform and change phenotype uncontrollably. It is usually thought that this phenomenon is of no significance, that it is an artefact connected with cultivating cells, and does not contribute any relevant information to differentiation produced *in vivo* in the organism. For the Darwinian model it assumes, in contrast, essential significance. It reveals the non-specific and probabilistic nature of cell differentiation. Having escaped from the selective constraint that the internal environment exerts over the organism *in vivo*, cells transform spontaneously owing to stochastic molecular interactions which are no longer channelled. In this context, what poses the problem is not so much explaining why two cells are different from each other, but rather understanding how, despite this inherent tendency to differentiate, homogeneous tissues of identical cells can form in the organism. The problem is not one of cell differentiation but of their 'identification'. To resolve this, what cell selection consists of must be more precisely explained.

6.3.2 *From metabolic selection to stabilisation by the 'signal-food'*

The increase in complexity of organisms arises not only from structural and functional innovation but also from their biochemical evolution which is characterised by the appearance of molecules with new properties. In ontophylogenesis, the origin of multicellularity lies in metabolic selection, but that obviously does not exclude the appearance of more sophisticated mechanisms in the course of evolution. Cell regulation in organisms existing nowadays is accomplished by molecules such as hormones or growth factors. These molecules have no nutritive value *per se* and seem to operate as signals. As with increase in the complexity of functions, this biochemical evolution does not invalidate the Darwinian model but improves it.

The instructive model comes up against non-specificity of the signals and leads to the same contradiction as genetic determinism. On the other hand, in a selective mechanism, the molecules which constitute the cell's microenvironment do not need to exert a specific effect since it is an adaptive mechanism, based on the stochastic behaviour of the cell. Sergei Atamas performed an analysis of the greatest importance on this subject. We know that antibody synthesis takes place according to a selective mechanism (see the introduction to this chapter). From a strictly formal point of view, there is therefore no difference between the function of a nutrient which allows selection of an animal in an ecosystem and that of an antigen which produces the same effect on a lymphocyte in the immune system. Of course, we spontaneously tend to think that these two systems have nothing in common, but if we think of them in a more abstract way the processes that take place are the same in both cases. A chemical substance, the nutrient or the antigen, permits the selective proliferation of an entity, the animal or the lymphocyte, and in both cases, the substance is broken down during interaction with the entity. Usually, the antigen is considered as a signal to the lymphocyte to proliferate and the nutrient just as the animal's food. This arbitrary way of creating classes, separating these chemical substances, depends on our subjectivity. It would be equally justified to consider the antigen as the lymphocyte's food and the nutrient as a signal to the animal to proliferate. In reality, neither of these stances is valid. The nutrition metaphor to describe these biochemical substances is no more relevant than the information metaphor. We must avoid falling into the trap of oversimplification. As biology is not a mathematical science, we have become trapped in the language. For a formal model representing these processes towards which scientific practice must move, it is necessary to create a new concept to describe what Atamas calls the 'signal-food' (1996, 2003).

To give an objective definition to the signal, the argument is often put forward that it exerts its effects at very low concentration and that it has no nutritive value in itself. However, all biochemical networks are interconnected in a cell. When a cell receives a signal (or a set of signals) which induces its differentiation and proliferation during embryogenesis, the chain of reactions participating in transduction of this signal must end in activating the cell's metabolism thus permitting the biosynthesis required. Leaving our subjective classifications aside, what happens here is a series of chemical reactions interacting one with another, and, in this respect, what we call signal transduction is nothing but extreme sophistication of the metabolism which, in the Darwinian model, permits the cell to adapt to its microenvironment. The signals are part of it in the same way that trophic substances are. We might then also call them 'selectors'. In the rest of this book, we shall, from habit, call them either signals or selectors.

6.3.3 *Role of signals in the Darwinian model*

The signal is always interpreted according to an instructive and deterministic model. If this model is wrong, what is its role, since in general it has no nutritive value *per se*?

The Darwinian model relies on selection of stochastic molecular events which change the properties of cells and let them adapt. The frequency of these events in this model is a key parameter. If it is very low, the cells will have a stable phenotype: their characteristics will not change. In contrast, if this frequency is high, the cells will be very unstable. Now it happens that this frequency itself depends on the stability of the molecular interactions.

Let us consider the simplest example possible. A single molecule of a regulator R can activate two genes a and b by binding to their promoter sites. When the molecule R has bonded to the promoter of a, it can escape by dissociating and diffusing as far as b in order to activate it (Fig. 21). The probability of this event of stochastic molecular recombination occurring depends on the stability of the molecule's bond with a. The more stable it is, the less dissociation there is and consequently the lower the probability of the molecule diffusing towards b. Conversely, the less stable the bond, the higher is this probability. It will be the same for all remodelling of the cell's protein complexes, whatever function they are involved in.

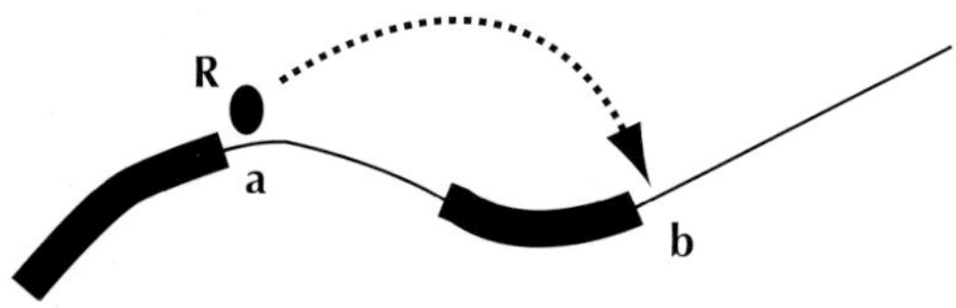

FIGURE 21. **Influence of the stability of random molecular interactions in gene expression.** The DNA wound around the spherical protein structures in the chromatin, called nucleosomes, is not expressed (thick line). This fibre has to be unwound to be accessible to gene transcription proteins (thin line). In these regions open to transcription, in order for a regulator R to be able to diffuse from gene a to gene b, it must first detach itself from a. How frequently this event occurs depends on the stability of the bond between R and a.

Looking at the situation in the context of the Darwinian model of differentiation, if the cell could adjust the stability of its state, i.e. the stability of the molecular interactions, depending on its needs, that would mean its functioning would have considerably developed; in other words, the cell would be in an unstable state when it is in an unfavourable microenvironment and conversely, in a stable state when in a favourable microenvironment. This would increase the probability of it changing phenotype and adapting to this microenvironment when necessary. In the example of the heap of cells (Fig. 16B), those in its upper part are in an environment poor in nutrients and their metabolism is slowed. It is therefore preferable for them to be in an unstable state which would allow them to change phenotype and adapt. On the other hand, the cells in the lower part of the colony are in an environment rich in nutrients and their metabolism is active. It is therefore preferable for them to be in a stable state and not to change phenotype.

In the Darwinian model, the function of signals is to induce these changes in stability depending on the needs of the cell (Kupiec, 1996, 1997).

6.3.4 *Mode of action of a signal* (*selector, signal-food*)

The stability of a cell state depends on the stability of the bonds between the molecules. This stability depends in turn on post-translational or epigenetic modifications of the proteins. For example, the phosphorylation state of proteins modifies their binding properties. This is a well established fact, especially as concerns gene expression regulatory factors binding with their target sequences in DNA (Li *et al.*, 1994; Xu *et al.*, 1994; Bourbon *et al.*, 1995; Lefebvre *et al.*, 1995; Takenaka *et al.*, 1995). Gene expression and the cell type that ensues from it are thus more or less stable depending on the state of phosphorylation of these proteins. This state of phosphorylation itself depends on signalling pathway enzymes activated by signals (see for example Hill and Treisman, 1995), so the latter have a direct influence on the stability of intermolecular bonds in the chromatin, and in this way control the stability of gene expression and cell type (Kupiec, 1996, 1997).

It is important to understand here that the action of the 'signal-food' in this selective context is non-specific. It only occurs to stabilise (or destabilise) a situation which in the first instance has been randomly produced, not to induce it in a deterministic mode.

Another thought experiment illustrates this in the context of the most rudimentary organism possible. This organism is composed of two cells each of which has a single chromosome carrying two genes *a* and *b* which can be activated stochastically by a single regulator molecule *R* as illustrated in Fig. 21. These two genes *a* and *b* code for the proteins *A* and *B*. These two proteins are signals localised in the cell membrane; their interaction activates a signalling pathway, within the cells, that leads to phosphorylation of the regulator *R*. The phosphorylation causes *R* to be stabilised on its binding site in the DNA. The stabilisation is non-specific. Regardless of whether the stabilisation concerns *R* in *a* or *b*, the same signalling pathway is always used. Once the germinal cell of this organism has divided into two cells, three random cell combinations are possible, *AA*, *BB* or *AB*. Out of these three possible outcomes, only *AB* leads to stabilisation of the cell types, owing to interaction between the proteins *A* and *B* and the subsequent stabilisation of *R* on the genes *a* and *b* in the two cells. The interaction of the signals therefore selects the combination *AB*. Until this combination is effected, the cell types are unstable. They change stochastically because the regulator *R* is not stabilised on the DNA. This mechanism thus forces the organism to evolve towards its stable 'adult' state (Fig. 22).

This stabilisation mechanism has been supported by experimental data which have shown that, as predicted, binding of the proteins giving structure to the chromatin, such as HP1 and histones, to DNA or to their protein partners, is very unstable in murine stem cells but is stabilised when these cells differentiate. These proteins control gene expression by determining the general organisation of the chromatin (Meshorer *et al.*, 2006).

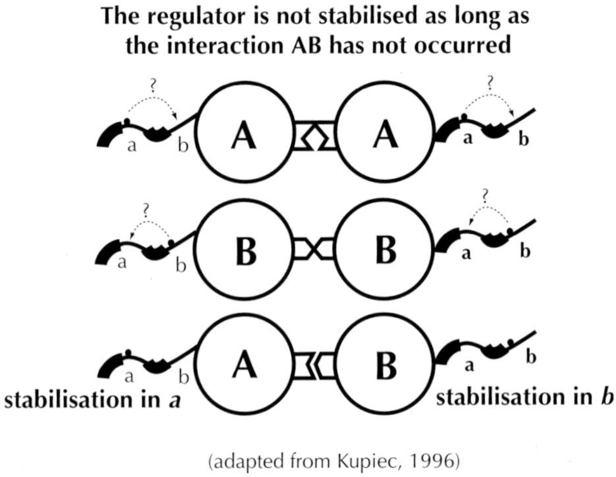

FIGURE 22. **Stabilisation of cell types by signals.** Only the interaction of the membrane proteins A and B corresponding to the simultaneous presence of R on genes a and b in two different cells triggers its phosphorylation and stabilisation on genes a and b respectively. The result is coordinated stabilisation of the cell types A and B.

Signalling pathways also lead to other epigenetic reactions such as acetylation and methylation. All the epigenetic modifications of proteins and DNA could therefore be involved in stabilisation by signals (Paldi, 2003).

6.3.5 *Experimental data relating to cell selection and stabilisation*

In the Darwinian model, interactions between cells are selective or stabilising, which agrees with many experimental observations. Embryogenesis of most organs is effected through differential multiplication of the cells, those relating to the organ being formed multiplying more quickly than neighbouring cells in the embryo. This has been demonstrated for organs as varied as those of the genito-urinary system, the nervous system, the heart, the liver, the ear, the limbs, the spinal cord etc. During embryogenesis, episodes of massive cell death occur which 'sculpt' the organs and are essential to its normal development (Michaelson, 1993; Penaloza *et al.*, 2006). Conceived conventionally, this cell death is said to be dictated by the genetic programme, with programmes existing to control it. However, this theory comes up against the same problem as cell differentiation. How can these programmes function with non-specific molecules? In the context of a deterministic view of the cell based on specific signals leading to very precise regulation, the very existence of this cell death is problematical. Why would the genetic programme create cells and subsequently destroy them? If cell differentiation is a selective process, on the other hand, it is normal for cells to die from the moment when they are not adapted to their microenvironment or if there are too many of them relative to the resources available (Kupiec, 1986; Glisse *et al.*, in press).

Several observations corroborate this interpretation. Growth factors very frequently act as trophic factors necessary for survival. They do not stimulate multiplication of the cells but their presence is essential to avoid the death of cells. During embryogenesis of the brain of vertebrates, 20 to 80% of the neural cells die (Gordon, 1995). A major proportion of this cell death is connected with the rarity of growth factors, only the neurons that have an adequate quantity of them survive (Vyas *et al.*, 2002).

Experiments performed on the development of the wing of the fly *Drosophila melanogaster* confirm the Darwinian explanation of cell death. In flies with two types of genetically different cells, some metabolically more active than others, the cells compete during development of the wing to avoid cell death. This competition is related to a 'decapentaplegic' survival factor and is an integral part of the embryogenesis process of the wing. This factor is usually considered a signal but for the Darwinian mechanism it really acts as a resource. In this competition, the cells with the more active metabolism monopolise the decapentaplegic factor, proliferate more rapidly and take it up to the detriment of the less active cells, which die (Moreno *et al.*, 2002; Moreno and Basler, 2004; Diaz and Moreno, 2005).

The stabilising role of signals is equally supported by experimental data. Wnt proteins are growth factors which play a major role during embryonic development (Logan and Nusse, 2004). In the

Drosophila embryo, expression of the gene *engrailed* depends on the protein Wingless (the equivalent of Wnt in this organism). In the absence of Wingless, *engrailed* can still be expressed initially but this is soon interrupted instead of being maintained as is the case when Wingless is present. The result has been confirmed by a series of further observations of similar significance (Arias and Hayward, 2006). In line with the Darwinian model, these data show that Wnt proteins do not play the role of inducer but rather of stabiliser in the stochastic expression of genes (Kupiec, 1996).

Major mortality occurs among blood cells, which are continuously replaced by cells that differentiate from stem cells. This haematopoiesis is regulated by a large number of proteins called cytokines which transduce their signal within cells *via* enzymes known as Janus kinases (JAKs) and gene transcription regulators. There are more than 50 cytokines for just four JAKs and seven transcriptional regulators. This system thus provides a good example of the non-specificity of molecules which underlies cell differentiation. We have already mentioned the experiment in which the erythropoietin (EPO) receptor was replaced by that of prolactin, resulting in blood cells differentiating normally under the action of prolactin, which usually acts in a different cell line (chapter 4, §4.3.2). A series of similar experiments has been performed with other cytokines and other haematopoietic cells and confirms that it is the state of the cell which determines its differentiation and not the chemical nature of the signal it receives. Furthermore, differentiation of haematopoietic cells can also take place normally if any cytokine deficiency is compensated for by the expression of a survival factor (bcl-2). These experiments indicate that the function of cytokines is not to specifically induce differentiation but simply to allow the survival and multiplication of cells (Robb, 2007).

6.3.6 *Testable predictions of the Darwinian model*

A theoretical model should propose predictions that can be tested and initiate a research programme, which is the case with the

Darwinian model. If the variability of gene expression is an essential biological parameter at the root of cell differentiation, and not simply noise due to the way a deterministic genetic programme functions, it should be possible to correlate the two phenomena, and that correlation should be compatible with the Darwinian model. Indeed, the latter predicts that during cell differentiation the variability of expression between cells itself must vary quantitatively. This variability may be expected to increase as long as the cells are not adapted to their microenvironment, because the genetic expression is not stable, and in contrast, decrease when the cells differentiate to adapt to this microenvironment, since a particular profile of genetic expression is selected and amplified corresponding to the cell type selected. No such evolution in variability is expected if the latter is the result of experimental noise independent of the physiological state of the cells (Heams, 2004). We can make an additional prediction concerning the molecular basis for this restriction in variability: as we have indicated, in the Darwinian model, signal transduction controls the variability of genetic expression *via* protein phosphorylation. This phosphorylation can be experimentally altered by various means, aiming, for example, at inhibiting or activating, even over-activating, the phosphorylation and/or dephosphorylation enzymes. This treatment should also therefore alter the restriction in variability of expression during cell differentiation and disturb its occurrence.

The first of these two predictions has started to be validated. When human embryonic cells cultured *ex vivo* multiply until they saturate the culture dish, they undergo physiological transformation called contact inhibition which causes them to cease multiplying. The variability of gene expression increases in this phenomenon in the first instance when the cells arrive at saturation and certain cells of the population are stabilised with a new gene expression profile (Neildez-Nguyen *et al.*, 2008; Stockholm *et al.*, in press).

Thanks to the collaboration of several researchers with complementary skills, these predictions of the Darwinian model have been greatly developed and are the subject of a specific experimental research programme (Guillaume Beslon, INSA-Lyon; Olivier

Gandrillon, CNRS, UCB-Lyon; Jean-Jacques Kupiec, INSERM, ENS-Paris; Andras Paldi, EPHE, Généthon-Evry).

6.4 Simulation of the Darwinian model

Computer simulation is another tool with which the validity of a theoretical model can be tested. We used this approach and created a computer program which brings into play virtual cells subjected to the rules of the Darwinian model (Laforge *et al.*, 2005; Glisse *et al.*, in press). This work was done in cooperation with physicists of the Pierre and Marie Curie University (Paris). Before describing our simulations, we need to explain a point of methodology.

Modelling does not consist in reproducing all aspects of reality, but is bound to be a simplification and abstraction of it. Otherwise, it is not modelling. Let us take the example of cell modelling which concerns us here. If we were to create a computer program which reproduced all the aspects and all the details of a real cell, it would no longer be a model but a virtual copy of the cell. Since it would be as complex as the original, this copy would be just as difficult to analyse, and would therefore be of little use. A model should only seek to capture one aspect of the reality and help us understand the contribution of this aspect to the real phenomenon which is necessarily always more complex. In our simulations, our aim was not therefore to mimic a particular situation by precisely describing all the parameters of a cell, but to analyse the general properties of the Darwinian model. The question we asked was as follows. If in a real complex cell there is a Darwinian mechanism based on chance and selection, what is this mechanism's contribution to the behaviour of the cell? More specifically we wanted to know whether it is in a position to create organised tissues reproducibly despite its probabilistic nature. We also wanted possibly to bring to light its non-trivial properties. Our simulations are, to some extent, thought experiments assisted by computer, allowing us to explore the model by going beyond what it is possible to do simply by reflecting. To perform them, we constructed the very simplest idealised cell model. The results we obtained show that it had certain properties essential to embryogenesis.

6.4.1 *Interstabilisation and autostabilisation produce different effects*

In the first instance, we produced a simulation with which we could only test the effect of cell stabilisation. In this simplified version, two cell types, RED and GREEN, fill a two-dimensional matrix. Each box of the matrix represents one cell which may be either of the two types, its identity being chosen by chance. At each simulation step, the cell can change type randomly. This is the probabilistic component of the model. However, the probability of its changing type depends on its environment which itself depends on the activity of the other cells of the population. The RED type of cells synthesises red molecules and the GREEN type of cells synthesises green molecules, these molecules diffusing in the matrix. At any moment therefore, a cell is in an environment characterised by concentrations of these red and green molecules.

The value for the probability (p), between 0 and 1, that a cell will change type is a function of these concentrations of red and green molecules. When $p = 0$, the cell is stabilised. This is the stabilising component of the model (Fig. 23). We tested two possible methods of stabilisation. Interstabilisation is stabilisation of a cell type by the molecules produced by the other cell type. In this case, cells of the RED type are stabilised ($p = 0$) when there is a large concentration of green molecules in their environment (when they exceed a certain threshold) and, in parallel, cells of the GREEN type are stabilised ($p = 0$) above a threshold concentration of red molecules. Autostabilisation is stabilisation of a cell type by the molecules which it synthesises. Here, RED is stabilised by red molecules and, likewise, GREEN by green molecules.

Figure 24 shows a typical result obtained for each method of stabilisation. At the start of these simulations, the matrix is filled with cells the type of which is selected at random ($p = ½$ for each cell

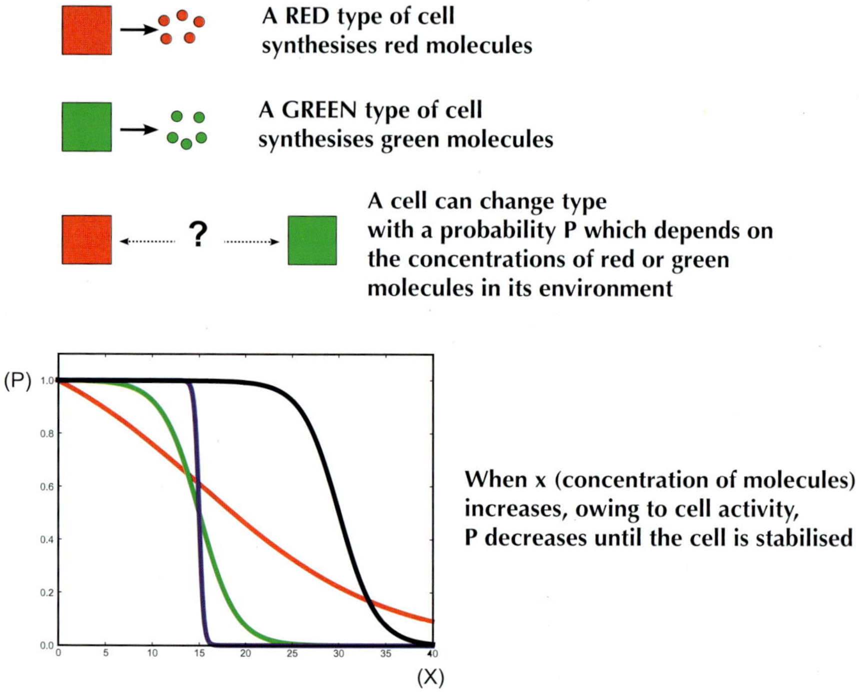

Figure 23. Characteristics of simulations of the Darwinian model of differentiation. The probability p for a cell of changing type at a given moment is a function F(X), X being the concentration of molecules in the cell environment at this moment. This function depends on two parameters β and C_0. β allows the slope of the function to be varied (examples: the blue, green and red graphs are obtained for various values of β with C_0 constant). C_0 allows the value X to be changed for p ≈ ½ (examples: the green and black graphs are obtained for two different values of C_0, with β constant). Depending on the values of β and C_0, the cells are stabilised at different molecular concentrations. The simulations are also controlled by a series of parameters the values of which may be changed: rate of synthesis of the molecules, rate of degradation, speed of diffusion of the molecules. During a simulation, cells stabilise in a type when the concentrations of red and green molecules increase. The aim of the simulation is to find out whether the states of equilibrium generated for the system produce organised structures with properties relevant for a biological system.

type), and the simulation continues according to the rules defined until the population of cells is possibly completely stabilised.

In both cases the population evolves towards a stable state, but as can be seen, interstabilisation and autostabilisation exert very

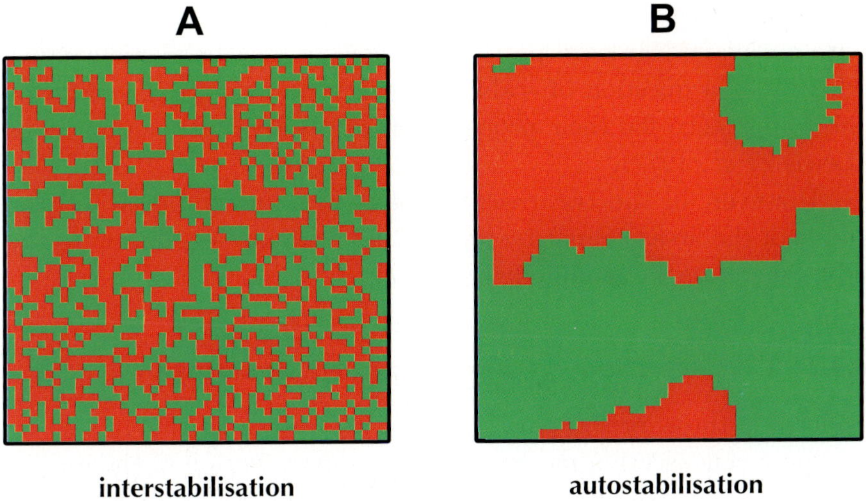

FIGURE 24. Inter and autostabilisation. Interstabilisation is stabilisation of a cell type by the molecules of the other type (e.g. RED type by green molecules). Autostabilisation is stabilisation of a type by its own molecules (e.g. RED type by red molecules). In both cases the cell system stabilises but interstabilisation produces small areas of cells while autostabilisation produces large ones.

different effects. In the case of interstabilisation (Fig. 24A), small areas of cells of each type are intertwined in the matrix whereas with autostabilisation, large homogeneous areas are produced (Fig. 24B). If the experiment is repeated several times, the cell population always becomes stabilised, with interstabilisation always producing small areas and autostabilisation always producing large areas. This difference between interstabilisation and autostabilisation can be easily understood. In the case of interstabilisation, one cell will promote the stabilisation of cells of the other type in its environment, which produces small areas. In the case of autostabilisation, one cell will promote stabilisation of cells of its own type, which produces large areas. This is reminiscent of lateral inhibition and lateral induction phenomena during embryogenesis. Lateral inhibition occurs when a cell inhibits the formation of adjacent cells of the same type, whereas lateral induction is the reverse. In the former case, mosaic tissues are formed from different cells alternating, whereas in the latter, homogeneous tissues are produced (Lewis, 1998).

Autostabilisation is therefore necessary for producing tissues with a certain degree of spatial extension. In a probabilistic context where it is a question of stabilising cell types, this result confirms the importance for their propagation of positive retroaction loops (Lewis *et al.*, 1977).

However, in the two cases of inter- and autostabilisation, the shape of the areas cannot be reproduced from one simulation to the next, therefore we modified the model.

6.4.2 *Cell selection creates organised structures*

Following these preliminary results, we improved the model by adding a cell selection mechanism. In the first version of the model, the cells did not multiply nor did they die. In this second version, the cells function in line with the autostabilisation model so that a certain expanse of tissues is obtained, and in addition, they are interdependent for survival and multiplication. This means that, to survive and be able to divide, a cell must metabolise a certain quantity of molecules produced by the other cell type. The red and green molecules thus have pleiotropic effects. They are involved in the autostabilisation of their original cell type and in selection of the heterologous cell type. They are analogous in this respect to real growth factors which are proliferation, survival, or differentiation factors depending on the cell context (see for example: Fortunel *et al.*, 2000; Tjwa *et al.*, 2003).

To sum up, therefore, in this improved model, in each RED or GREEN type of cell and at each stage of simulation:

(a) red or green molecules are synthesised;
(b) some of the molecules are broken down;
(c) the molecules not broken down diffuse;
(d) the identity of the cell is determined by a law of probability which is a function of the red or green molecular concentrations, and is identical to that used in the first version of the model and systematically used depending on the mode of autostabilisation;

(e) if the cell is RED, it metabolises green molecules. If it is GREEN it metabolises red molecules. Depending on the concentration of these molecules, the cell either dies, if there are not enough, divides if it can metabolise enough of them, or stays alive without dividing until the next simulation step. This is the aspect of the model that corresponds to cell selection. The only cells that survive and proliferate are those of a type adapted to their microenvironment determined by the concentrations of red and green molecules.

The results obtained with this model are very much more conclusive than those obtained with the first version without cell selection. An organised tissue structure is produced in a way which is reproducible.

Figure 25 shows the typical result of a simulation. To start with, the matrix is seeded with 16 cells the types of which are chosen at random (Fig. 25A). The growth of these cells gives rise to a longitudinal structure composed of two cell layers, RED and GREEN (Fig. 25B). The limit between these two layers is well defined throughout the length of the structure and the thickness of the layers is regular. This bilayer of cells continues to grow longitudinally (Fig. 25C) until it reaches an 'adult' state when growth ceases (Fig. 25D). From this moment, the simulation may be allowed to continue but the cells will not multiply further and the bilayer of cells will remain as it is. This structure possesses an invariable characteristic — its organisation in a bilayer, and a variable characteristic — its longitudinal shape. In other words, if we perform other simulations starting with 16 new cells selected at random, with identical parameter values, the bilayer created will always have the same characteristics (thickness of the layers of RED and GREEN cells) but the longitudinal form of it will be different (see for example the structure of Fig. 26A). However, in some simulations, the bilayer does not form and if this is the case, in general all the cells die. More rarely, they remain in a disordered state. This mortality is compatible with experimental reality. In all species there is a considerable level of embryonic mortality which is in large part

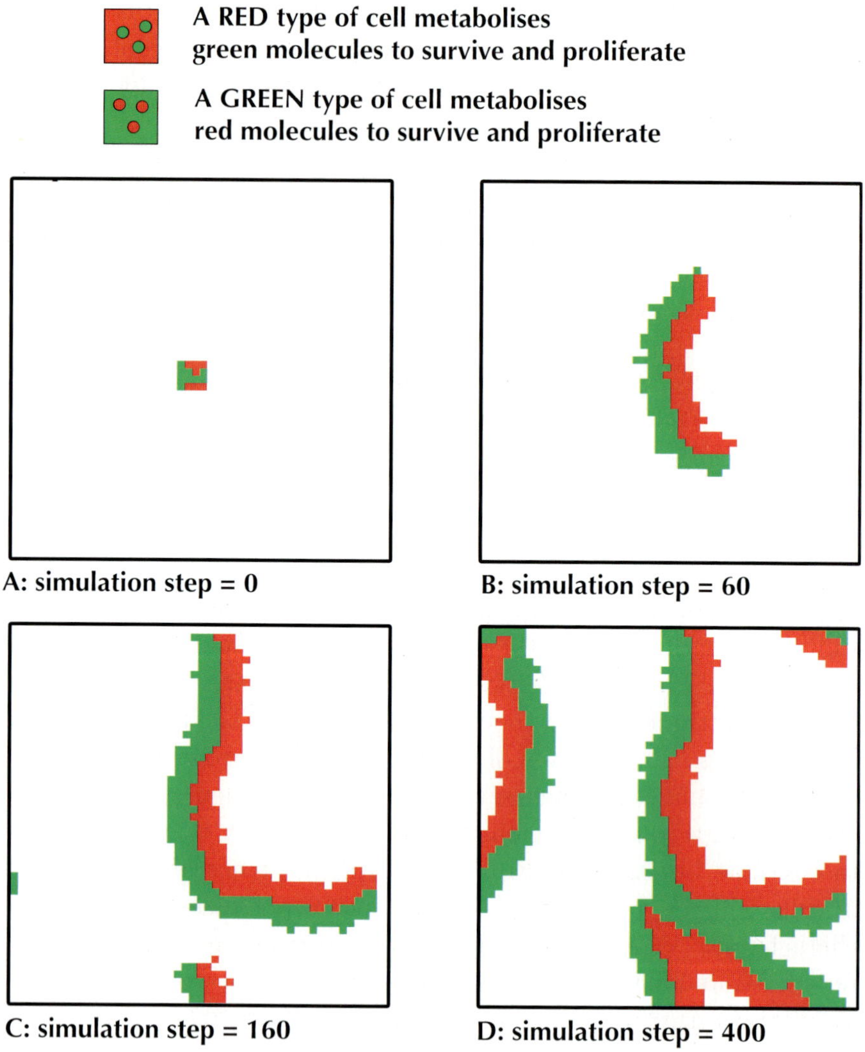

Figure 25. Formation of an organised tissue structure. The principles of the previous model are retained (the autostabilisation version) but cell selection is added in the form of interdependence for survival and proliferation. A cell must metabolise a minimal quantity of molecules produced by the other cell type. This selective model involves new parameters — quantities of molecules necessary for survival and cell division. It induces the formation of an organised tissue structure as a bilayer of cells.

unexplained but is predicted by a probabilistic theory of embryogenesis (Kupiec, 1983). In our computer model, this rate of failure can be reduced to less than 10% by optimising the values of the parameters.

If we perform a large number of simulations, we create the equivalent of a virtual species: a series of structures forms which are different from each other even though their tissue organisation is common to them all. This phenomenon is similar to what happens in a biological species in which all the individuals are different even though they share the same organisation. In certain species, such as those of mammals, variability from one individual to another is relatively limited. In other species, such as certain plants, it can be much greater. Our virtual species is obviously very simple and scarcely constrained compared with a real species, so its longitudinal variability is great.

The mechanism brought into play here is different from a reaction-diffusion or self-organisation phenomenon. In these theories, the change of state of a system depends on specific reactions between its components and on differences in their diffusion speed. Randomness is only involved as a fluctuation setting off the deterministic dynamics which make the system swing from one state of equilibrium to another. From the moment the dynamics are set in motion, there will always be a change in the state of the system (see Fig. 3). In contrast, the Darwinian model involves an intrinsically probabilistic mechanism based on non-specific reactions (cells changing type according to probabilistic laws) subject to a selective constraint (interdependence for proliferation). Due to the intrinsically probabilistic character of this model, from the moment the proliferation of cells is triggered there is always a certain rate of failure in their forming a cell bilayer structure. In addition, even though the values of the parameters of diffusion are identical for the two types of molecule (which is the case for the examples we are presenting) the system is being structured. This structuring does not depend therefore on a difference in diffusion, as a reaction-diffusion model implies.

6.4.3 *Spontaneous growth arrest is the result of equilibrium between cell selection and phenotype autostabilisation*

Not only does simulation of the model generate a reproducible cell bilayer, but this structure also spontaneously stops growing, as does a real adult organism. This is not a trivial property. There is no rule specifying possible cessation of the multiplication of cells in the computer program controlling the simulations. This is a property intrinsic to the Darwinian model, generated by the way it functions.

The arrest of longitudinal growth can be easily understood. Due to the toroidal structure of the space represented by the matrix, the two ends of a bilayer always end in meeting to form a loop which prevents longitudinal growth. On the other hand, the arrest of lateral growth is much more remarkable. To multiply, a cell needs molecules produced by the other type of cell. Since these molecules diffuse, they form concentration gradients decreasing as they move away from their source. This means that the concentration of red molecules decreases starting from the RED layer of cells as do the green molecules similarly, but from the GREEN layer of cells. We have analysed the concentrations of molecules in cross-sections of the bilayer and have checked that these gradients do indeed exist. It is therefore normal for the cells to stop dividing in the external zones of the bilayer through lack of molecules. The RED type of cells situated in the external zone of the RED layer do not have access to a sufficient number of green molecules to be able to multiply and, similarly, the external GREEN type of cells lack red molecules. However, on the outside of the RED layer there are sufficient red molecules for the GREEN type of cells to multiply there and likewise, on the outside of the GREEN layer, there are sufficient green molecules for the RED type of cells to proliferate — but that is not what happens. The explanation for this is that in these zones of the bilayer, due to high concentrations of red and green molecules, the RED and GREEN cell types are completely autostabilised (the probability of a cell changing type is zero), which inhibits the formation of cells of the other type which could proliferate. This means that cells ceasing to multiply is the result of the joint action of mechanisms of interdependence for proliferation, i.e. selection and autostabilisation. Interdependence prevents the multiplication of cells that lack substrate because they are too far away from the source of it, whilst autostabilisation fixes these cells as the type they have acquired and prevents growth of the other cell type.

Through this combined action of the two mechanisms, continuation of overall growth of the structure is inhibited. We have checked the relevance of this analysis by alternately deleting autostabilisation and interdependence of cells for proliferation from the model. In both situations total loss of the properties of organisation is observed. Cells no longer cease proliferating spontaneously. This produces infinitely alternating areas of RED and GREEN cells which are only restricted by the size of the matrix (Fig. 27C). In as far as the two processes of interdependence and autostabilisation themselves depend on the quantitative value of all the parameters of the model, growth arrest is the result of equilibrium between the values of these parameters. If we distort these values one by one, it is possible to induce a loss of organisation properties in the bilayer. Figure 27A shows the example of distortion of the speed of diffusion. Here, the cell population is again growing infinitely with cells of the RED and GREEN types overlapping.

6.4.4 *A new conception of cancer*

The results of the simulations suggest cancer could be seen in a new light. In conventional theory, the genetic programme not only regulates the differentiation of cells through the signals it emits but also their proliferation. The multiplication of a cell is activated or inhibited by appropriate signals. In the Darwinian model, the functioning logic is quite different. There is no difference between the cell system in a state of growth and the quiescent system which might be connected with the action of a signal controlling multiplication. In the simulations, the cells cease to proliferate spontaneously without any functioning rule for the model stipulating this.

When growth stops, it is quite simply because the system has reached a state of equilibrium in respect of the quantitative values of its parameters. Disruption to the overall equilibrium of the system leads to proliferation being resumed.

To illustrate this we performed another simulation experiment. In the first instance, we allowed a bilayer of stabilised cells to form until it stopped growing (Fig. 26A). Then we modified the value of one parameter involved in cell autostabilisation (the parameter C_0 of the function that determines the probability of a cell changing its type; see the caption for Fig. 23). We progressed gradually until we found the limit which destabilises the structure without destroying it completely. As can be seen in Fig. 26, we found ourselves looking at localised limited resumption of proliferation, by cell budding from the bilayer. This resumption of proliferation, which can be produced at different locations on the bilayer depending on its shape, creates cell masses which are released into the environment (Figs. 26B, C, D). Similar results can be obtained if other parameters are modified, for example, those affecting the quantities of molecules present in the cell environment.

This simulation experiment suggests that cancer may arise from such disruption to the equilibrium between the selective action of the microenvironment and that of autostabilisation of the cell types, i.e. the stochastic expression of the underlying genes on which the synthesis of the molecules *a* and *b* in the model implicitly depends. In an actual organism, the action of a carcinogenic agent which fixes on a protein could modify its properties of diffusion into the tissues or its fixing on the DNA. Modifications such as these would lead in turn to upsetting the concentrations of this molecule in the tissues or destabilisation of the expression of genes, resulting in imbalance permitting cell proliferation to resume. This conception of control of proliferation does not mean that DNA mutation has no role to play *per se*. Such mutation could produce identical effects by modifying the properties of a protein.

Obviously a real organism is much more complicated than our cell bilayer and disruption of the equilibrium of this nature could

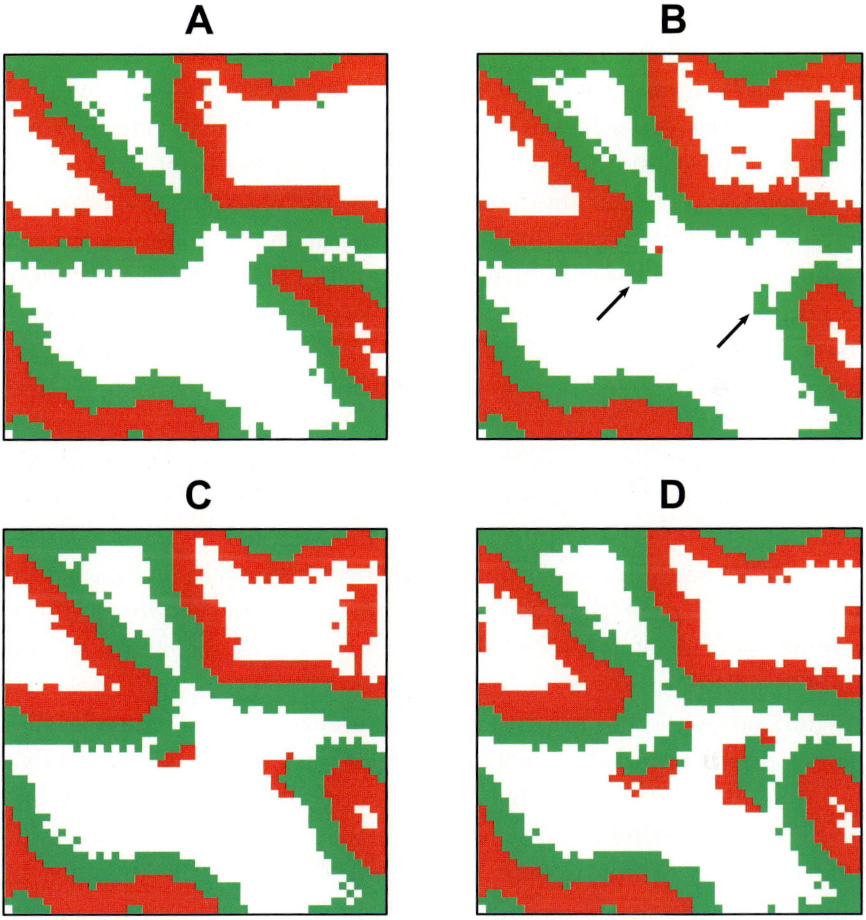

Figure 26. Simulation of carcinogenesis. A: A bilayer of cells is in a quiescent state. **B:** The cell types are destabilised by modification of the value of the parameter C_0: localised cell proliferation resumes. **C and D:** This resumption of proliferation gives rise to the release of cell masses into the environment.

arise from modification of a multitude of factors. We simply want to suggest that cancer could occur from a similar logic and not from alteration in the way the genetic programme functions. There is moreover already a series of experimental arguments in favour of this conception. Several observations back the role of the cell microenvironment. Carlos Sonnenschein and Ana Soto have put forward a theory in which cancer is not caused by mutations but by disturbances arising in tissue organisation and not at molecular level (Sonnenschein and Soto, 1999). This theory seems to be supported by tissue graft experiments demonstrating that N-nitrosomethylurea does not induce mutations in the cancerous tissue itself to exert its carcinogenic effect. In these experiments, the carcinogenesis mechanism seems to work through deregulation in the tissue adjacent to the cancerous tissue and not through the direct action of mutation in its cells (Maffini *et al.*, 2004). Mina Bissell and her team have likewise demonstrated that cancerous cells can become normal again when they are treated by signalling pathway inhibitors (Kenny and Bissell, 2003). These data put the role of mutation into perspective and show that the cancerous transformation of a cell depends on its environment (van Kempen *et al.*, 2003; Bissell *et al.*, 1999). They can be integrated into the Darwinian model because it stipulates that tissue organisation depends on the microenvironmental selective constraint that it exerts on the cells. The model can even remove inconsistencies between experimental data which are at first sight contradictory.

Indeed, while the role of the microenvironment in cancerisation is now a well-established fact, the role of mutation is equally undisputable. How can the two be reconciled? Jean-Pascal Capp has made an interesting analysis of this subject. We have already indicated that according to our model mutation could be involved in destroying overall tissue equilibrium, but this equilibrium could also be disrupted due to disturbances in the microenvironment. In both cases, the cells would be subject to gene expression deregulation since this expression is controlled by the microenvironment. Deregulation could then affect the genes, which are known to control rates of mutation, and cause an increase in the latter in the cells. In turn, through their effects these mutations would contribute to extending cell cancerisation. In this context, mutation is not therefore the obligatory initial cause of cancer but an effect of the imbalance between the influence of the microenvironment and the stochastic expression of genes. It is an aggravating factor in the process of cancerisation (Capp, 2005, in press).

6.4.5 *The role of morphogenetic gradients in the Darwinian model*

Morphogen molecules form gradients in the embryo (Gurdon and Bourillot, 2001; Tabata and Takei, 2004). In deterministic theories, it is acknowledged that cells differentiate specifically according to their position in these gradients, i.e. according to the concentration of morphogens in their immediate environment (Wolpert, 1989). In order to analyse the role of gradients in the Darwinian model we observed how they were formed at the start of the simulations, when the bilayer was being constituted. Their formation is simultaneous with that of the bilayer: they are the result of the same dynamics, as the gradients and the cell bilayer gradually and mutually reinforce each other. The bilayer forms owing to stochastic changes in the cell types, but this process can only occur because it is stabilised by the parallel formation of the molecular gradients. At the same time, the cells differentiate according to their position in the gradients. Cells of the RED type are found where there are most red molecules and in the same way, cells of the GREEN type coincide with the peak of the green molecule concentration. There is therefore no contradiction between the Darwinian model and the existence of morphogenetic gradients.

6.4.6 *Does the Darwinian model lead to the emergence of new properties?*

The relevance of the concept of emergence that we have already questioned (chapter 5) can be re-assessed by simulating the Darwinian model. Can we consider cessation of the multiplication of cells in the bilayer, which is a spontaneous non-programmed phenomenon, as an example of emergence? Growth arrest would seem to be a property at the tissue level of the system, emerging from interactions between cells. To answer this question we must remember that the concept of emergence presupposes a fundamental plan explaining the origin of order: the interactions between the components of a system brought into each other's presence induce new non-predictable properties which appear suddenly and spontaneously without the involvement of a cause external to the system itself. The system becomes qualitatively and irreversibly different from what it was before. There is thus a discontinuous jump from one level of organisation to another and the rules of a higher level are irreducible to the rules of the lower level (inexplicable using the rules of the lower level).

Simulations of the Darwinian model do not confirm this emergentist assertion, quite the reverse (Kupiec, 2005). Interpreting the spontaneous cessation of cell multiplication in our simulations as an emergent property of this nature is totally mistaken.

Figure 27 shows the results of five simulations. All these cell populations were produced by the same model that produced the

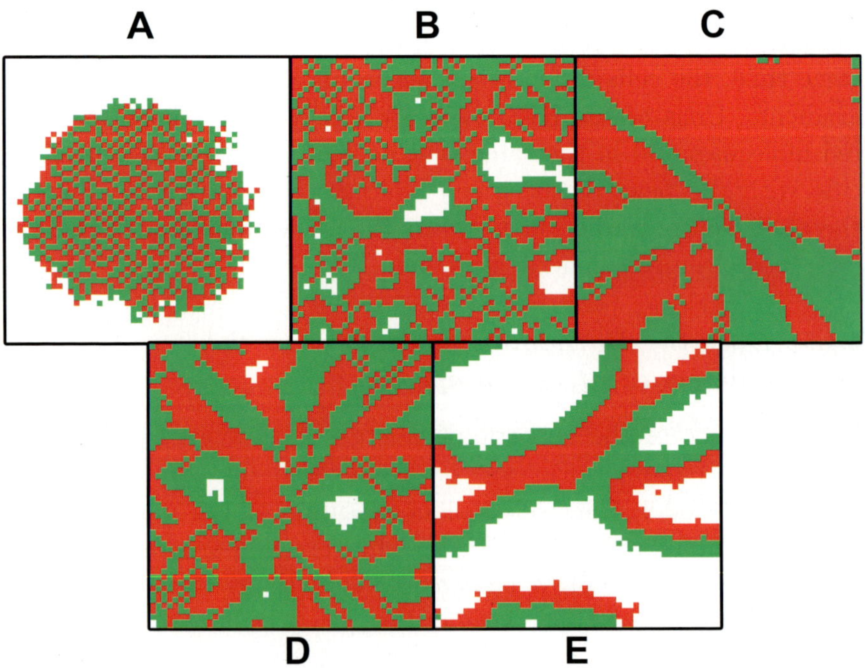

Figure 27. Organisation gradient. Quantitative variation in the parameters of the model causes the gradual appearance of the bilayer of cells, yet the model always remains the same qualitatively. There is no discontinuity or irreversibility in setting up this 'organisation gradient'.

bilayer. In each case the rules applying to the cells were strictly identical, the only differences being in the quantitative values of the parameters. As we can see, there is gradual movement from a totally disordered situation to the organised bilayer. There is no discontinuous break corresponding to the appearance of a new property which might qualitatively change the system. The order corresponding to the bilayer is not, by nature, different from the other disordered cell configurations. In addition, experiments simulating carcinogenesis (Fig. 26) show that order is not irreversible as would have to be the case if it corresponded to the emergence of a new property. It only needs a minor quantitative modification of a parameter for it to be destroyed. In fact, we are dealing here with a counter-example which shows to what point the concept of emergence is a trap: the simulation illustrates how a property we had not predicted and which seems to appear spontaneously in tissues can be perfectly explained and reduced to the rules regulating how cells function.

6.4.7 *Is the body a cell ecosystem?*

The simulation experiments that we have performed demonstrate that the Darwinian model has essential properties that one would expect of a theory of cell differentiation, such as the ability to create an organised structure, finite growth and reproducibility. Without demonstrating *per se* that real organisms comply with this model, this proves that it is a relevant general theoretical context for analysing embryogenesis and interpreting experimental data.

Simulations can also be used to evaluate the theory of the body as a self-organised cell ecosystem, as suggested by Jim Michaelson (1993) then considered by Pierre Sonigo in collaboration with Isabelle Stengers (Sonigo and Stengers, 2003). According to Michaelson, the relationships between the different types of cells of an organism can be compared to the instances of prey/predator equilibrium that are established in a self-organised ecosystem. An imbalance in these relationships is consequently the source of the diseases which may affect it. This theory should not be confused with ontophylogenesis. For Michaelson, embryogenesis is indeed governed by a selective mechanism, but this selection is purely internal to the organism. It does not reflect the external selective constraint which is exerted on the embryo, as is the case in ontophylogenesis (see Figs. 16 and 17).

We believe that the metaphor of the ecosystem can be used to describe the relationships concerning metabolic exchange between the cells[39] but, like any metaphor, it should be used sparingly because it may be dangerous. On the one hand it helps in understanding an aspect of the phenomenon, but on the other it induces misinterpretations. Embryogenesis cannot be reduced to this. Ontophylogenesis is a process not of self-organisation but of hetero-organisation, in which environmental constraints are essential. In addition, in this process DNA also retains a primordial role for which there is no equivalent in an ecosystem. It does not just passively provide the proteins which the cell needs. The way it functions, although probabilistic, involves rules allowing it to have an influence on the organism (see the following sections of this chapter). This point of view is moreover supported by the result of the simulations. When autostabilisation is eliminated and cells are reduced to their metabolic exchange relationships, as occurs in an ecosystem, they lose their properties for organisation (Fig. 27C). That does indeed suggest that biological organisation depends on equilibrium, but it is not a prey/predator (ecological) type of equilibrium between the parts of the organism. It is equilibrium between the equivalent of two forces: the pressure of selection that is exerted on biological structures, and the random non-specificity of molecular interactions which makes them variable.

6.5 Models of gene expression

If, as we believe, DNA plays an important role without it being that of a genetic programme, we must define it more exactly, which we shall now do.

---

[39] This metaphor is in fact similar to that of the organism considered as an economic system, used by Bernard right from the 19th century (see this chapter, §6.1.3), of which there are very many variants.

6.5.1 *Networks with noise*

The stochastic expression of genes is a phenomenon that has been undeniably demonstrated. However a great many researchers are divided in their adherence to two different interpretations. The first interpretation takes into account the properties of networks of biochemical reactions while the second ensues from analysis of the structure of chromatin. Neither of these conceptions accounts for the whole phenomenon. According to the first interpretation, stochastic expression is caused by disturbances which affect the way gene networks function. We know that between two cells there are always inevitable little differences in the concentration of molecules and in the speeds of chemical reactions. These fluctuations may have considerable consequences if they affect molecules present in low concentrations, the activity threshold of which is itself fairly low. The probability of exceeding the activity threshold in certain cells and not in others is all the greater the lower the activity threshold of the molecule relative to the fluctuations in concentration of that molecule. For a transcription factor which has to activate genes, the transcription noise which results from the fluctuations can lead to the stochastic expression of these genes and produce heterogeneity of types in a population of cells which were identical at the start.

This differentiation will be all the easier if there are regulatory loops in the gene network (Fig. 28; McAdams and Arkin, 1997, 1999; Kepler and Elston, 2001). This conception thus accepts stochastic expression without challenging the theory of genetic programming. It reduces randomness to noise that improves the way the programme works by permitting cell lines to bifurcate. Yet it still assumes that there is a microscopic network of intermolecular relationships determined by the stereospecificity of proteins, and that this network is an order underlying macroscopic cellular organisation. It does not therefore challenge the molecular biology principle of order from order. Here, the way the genetic programme functions is as a deterministic phenomenon with noise (chapter 2, §2.2.4). Now, we have seen that it is difficult to hold such a view given the generalised non-specificity which affects proteins, including

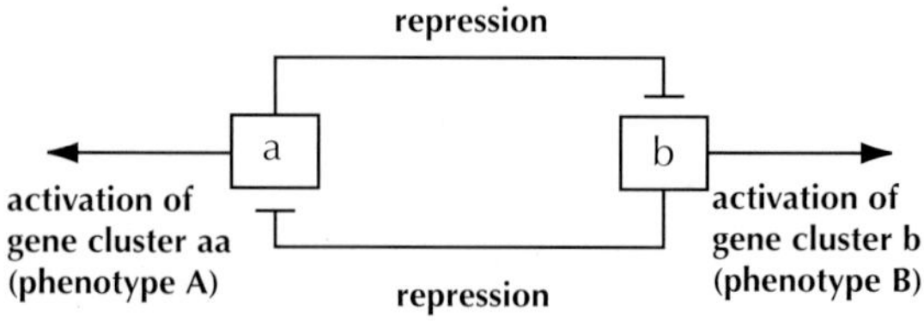

FIGURE 28. **Bistability of a network of genes.** Two genes mutually repress each other in the same cell. Gene *a* represses gene *b* and gene *b* represses gene *a*. Fluctuation in this network which increases the activity of one of these two genes to the detriment of the other augments until the latter gene is totally repressed and the former activated. As fluctuations are inevitable and can be randomly produced in favour of one or other of the two genes, some of the cells in a population will bifurcate towards type A (corresponding to activation of *a*) and others towards type B (corresponding to activation of *b*).

those involved in gene expression (chapter 4). This does not mean that all the work done in this theoretical context is wrong but that another, much more important source of randomness must be added to it, due to the combination possibilities ensuing from the non-specificity of chromatin molecules.

6.5.2 *Self-organisation model of chromatin*

According to the second interpretation of the stochastic expression of genes, the architecture of chromatin and the cell nucleus is thought to arise through a process of self-organisation involving random modifications of their structure. This process occurs spontaneously due to local interactions between the chromatin molecules, and tends towards the thermodynamic state of maximum equilibrium (Dundr and Misteli, 2001; Misteli, 2007; de Laat and Grosveld, 2007). Adherents to this conception try to justify it using a great deal of data obtained recently using the most sophisticated techniques. These data reveal cell nucleus properties which at first sight seem paradoxical. The nucleus is both extremely structured and extremely dynamic. Each chromosome is organised in a territory specific to it, which means that the genes are precisely positioned in the three-dimensional nuclear space. Genes co-expressed in one cell have then a statistical tendency to be co-localised in the same place in this three-dimensional space, in 'transcription factories' where the regulatory and transcription factors necessary are optimally concentrated (Cremer and Cremer, 2001; Chubb and Bickmore, 2003; Fraser and Bickmore, 2007; Misteli, 2007). However, while maintaining this overall architecture, chromatin is also extremely labile. It is continually assembled and disassembled by a flow of molecules which associate with and dissociate from each other.

The interactions between two proteins, or a protein and DNA, are very brief, for just a few seconds in the large majority of cases. After dissociation, the diffusion of these proteins in the nuclear space and their action on another site allows more or less important chromatin reorganisation to occur, leading to the expression of different genes. Owing to the intrinsically random nature of these processes, these modifications in the expression of the genes are themselves stochastic (Misteli, 2001). In reality, this is not prediction of a model of self-organisation here, but a consequence of the Brownian character of interactions between molecules in chromatin which is the basis of the stochastic expression of genes (Kupiec, 1983, 1989, 1996; see following section). An example is provided by the phenomenon of 'position effect variegation' of the expression of genes localised at the limit between the euchromatin and the heterochromatin. Chromatin is formed by an elementary fibre of DNA twining itself around nucleosomes which are themselves made of histones. This elementary fibre can wind around itself and form more compact structures. The two forms co-exist in the cell nuclei. The euchromatin corresponds to the loosely packed elementary fibre and the heterochromatin to the compact form. The genes in the latter are in general repressed because it is difficult for transcription factors to access them, whereas the genes in the euchromatin are expressed because they are more accessible. It was noticed a long time ago that genes localised at the limit between the euchromatin and heterochromatin are randomly expressed in a population of cells that is elsewhere homogeneous.

This phenomenon of variegated expression is due to the stochastic character of the molecular interactions which form the two

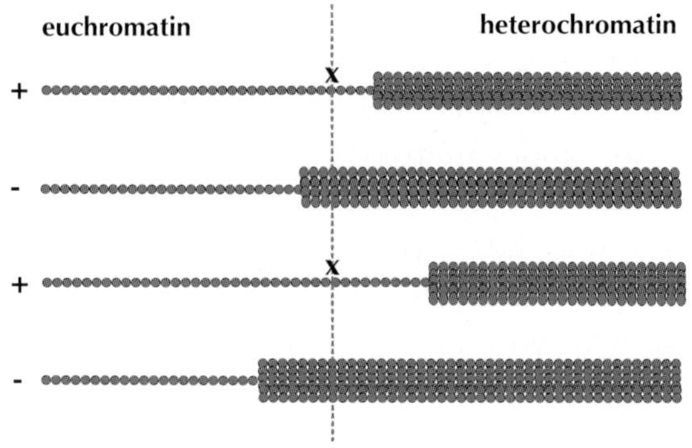

FIGURE 29. 'Variegated' gene expression. Owing to the intrinsically random nature of the interactions between proteins and DNA in the chromatin, the limit between the euchromatin and the heterochromatin varies. Gene x, located in this region, may be expressed or not depending whether it is in the euchromatin or the heterochromatin. The two cases are produced randomly in each individual cell of a tissue, leading to the 'mosaic' expression of this gene. Some cells in a tissue express it, others do not.

types of chromatin and transform the one into the other. These processes depend on competition between proteins promoting the formation of heterochromatin and transcription factors promoting euchromatin (Dillon and Festenstein, 2002). In each cell the result of this competition is random because the competition itself depends on the Brownian behaviour of molecules (Fig. 29).

However, while variegated expression illustrates a perfectly real phenomenon of stochastic gene expression, the model of self-organisation of chromatin suffers from the same defects as the general theory (chapter 5). In the course of embryonic development, in the huge majority of cases, cells do not form mosaics of differentiated cells but homogeneous tissues. If cell differentiation depends on self-organisation of chromatin, why this process leads to identical states in whole tissues, while being different depending on the tissues, needs to be explained. Stochastic expression explains why cells different one from the other are produced, but not the stabilisation of cell types which is necessary to form homogeneous tissues. Constraints must be exerted on stochastic expression in order to guide it, and that is indeed the case. Chromatin molecules are continually affected by signals (selectors) arising from the cell environment which cause epigenetic modifications (phosphorylation, acetylation, methylation etc.) that change the interaction properties of the molecules making it up (the diffusion coefficient, association constants etc.). Chromatin organisation depends directly on these parameters and is therefore controlled by these constraints arising from the cell environment. Here, again, we are dealing with a hetero-organisation process, not spontaneous self-organisation.

6.5.3 *The stochastic expression of genes subject to natural selection*

Two types of interdependent constraints are exerted on gene expression during embryo development: those that arise from its immediate development (its ontogenesis) and those that arise from its history (its phylogenesis). Firstly, selective constraints of the cell microenvironment are created in the embryo owing to its own development. They induce stabilisation (or destabilisation) of gene expression (Figs. 16 and 22). Secondly, embryonic development is itself constrained by its initial conditions, i.e. by the structure of the egg which is the product of its evolutionary history (Fig. 15). DNA plays a dominant role in the context of this phylogenetic constraint, because it is passed on in a (relatively) unaltered way[40] to each generation. By being unaltered it promotes the reproducibility of ontogenesis. Our theory does not therefore deny its importance in biological processes but attributes a different role to it. As for the synthetic theory of evolution, for our theory DNA is the result of external selective constraints to which the organism has been subjected and which have been interiorised in its structure *via* genetic mutations and recombination. In ontophylogenesis, however, it acts as a random protein generator, not as a genetic programme. The important point to emphasise in this conception is

[40] At each cell division, there is always a certain rate of genetic recombinations which modify the organisation of the genome.

that by its structure, by the relative position of the genes, DNA continues to have an effect on cell behaviour. It is not reduced to the role of a simple passive supplier of protein components or RNA that the cell might use as it sees fit. At the present time, a great many researchers are reconsidering genetic determinism. They include some who not so long ago were staunch supporters of it. In a kind of pendulum movement, they are coming round to denying any active role for DNA. We think that is an error.

The model of DNA as a random protein generator (Kupiec, 1983, 1989, 1996) is based on two properties of chromatin molecules: first, they move by Brownian diffusion to find their target sequences in the DNA (Berg and von Hippel, 1985; Halford and Marko, 2004), and secondly their interactions are not specific. Due to these two properties, the structure of DNA molecules determines gene expression probabilities during embryonic development. This model can be summarised by two general principles which explain both the structure of chromatin and gene expression.

First principle: The non-specificity of interactions between molecules leads to countless interactions between chromatin molecules[41] and their DNA binding sites. The result is a great many possible distributions of the molecules on the binding sites. Each of these distributions corresponds to a different chromatin structure and to activation of different sets of genes. Due to the diffusion of regulator elements and interactions between molecules being intrinsically random phenomena, the structure of the chromatin and gene expression are also random phenomena, each structure having a certain probability of being fulfilled.

Second principle: Because the interactions between chromatin molecules are unstable, the molecules can be randomly redistributed on their binding sequences, though the transitions between different distributions are not all equally probable. The probability of transition between two distributions depends on two main parameters: on the one hand, the stability of the interactions between the

[41] This principle remains valid whatever the chemical nature of these molecules.

chromatin molecules and their DNA binding sites, and on the other, the relative position of the genes in the DNA molecule. These parameters determine the chronological gene expression sequences which have the highest probability of occurring in a cell during embryogenesis and which correspond to the different cell differentiation pathways.

To illustrate these two principles, let us take the simplest example possible. The egg cell of an embryo has a single extremely rudimentary chromosome. It is made just of a single transcription regulator protein[42] (*R*) and a linear DNA carrying three genes G1, G2 and G3, positioned in this order with a distance d1 separating G1 and G2 which is greater than the distance d2 separating G2 and G3 (Fig. 30).

When *R* binds to the regulator region of one of these three genes, it activates it. In our egg cell, *R* is at G1 when embryogenesis begins. It is the result of the differentiation of the germinal line in the previous generation (Fig. 30A). G1 is activated as long as *R* is bonded to its regulator region. When it dissociates from it, it moves around randomly in the surrounding space (diffusion). Several events may then possibly occur. It may re-associate with G1, or escape from this space and, after translocation, bind with G2, or with G3, activating them in turn. These three events are random because the diffusion of *R* is itself a random phenomenon, but the probability of each occurring is not equal. Once *R* has dissociated from G1 (Fig. 30B), the probability of it re-associating with one of the three genes depends on their respective positions, since it explores the space randomly. This is a direct consequence of the laws of diffusion. It is most probable that it will re-associate with G1, therefore, and least probable that it will associate with G3, with the probability of it associating with G2 being intermediate. If in a cell the *R* molecule goes from G1 to G2 or G3, the same phenomenon is

[42] Here we are giving a wide meaning to the concept of transcription regulator that we are using. It means any molecule which influences gene expression including those participating in giving the chromatin its structure.

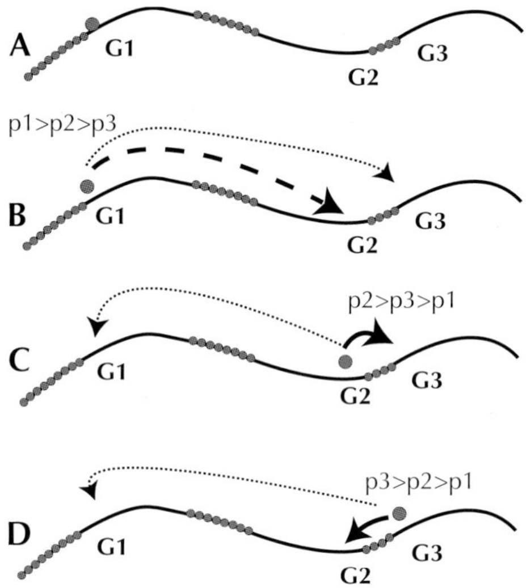

FIGURE 30. Model of stochastic gene expression. The regulator can interact to activate G1, G2 and G3, with the probabilities p_1, p_2 and p_3 respectively. These three probabilities are not equal. They depend on the position of the regulator and the relative distance separating the three genes.

reproduced. There is a probability of it translocating from its new position onto one of the two other genes depending on their relative positions. For example, when it is on G2, the highest probability is of it translocating onto G3 as the latter is closest (Fig. 30C). Similarly, when it is on G3, the highest probability is of it translocating onto G2 (Fig. 30D). The relative position of the genes determines the translocation sequences of *R*, and thus activation of the genes which have the greatest probability of being expressed. In our example, the sequence G1 → G2 → G3 is the most probable, but other sequences (G1 → G3 → G2, G1 → G2 → G1 etc.) are equally possible though less probable. In a given cell, these sequences determine the chronological sequence for expression of the genes. In a population of cells, the probabilities of expression of the genes associated with these sequences determine at a given moment the statistical frequencies of cells which express one of the three genes.

In this model, these characteristics of embryogenesis are therefore determined by the structure of the DNA which acts as a random protein generator and not like a deterministic genetic programme. Natural selection optimises the way this model functions by sorting the genetic recombinations, which are produced in each generation and change the relative position of the genes so that the probabilities of activation sequences corresponding to cell lineages of the embryo are themselves optimised and the overall viability of the organism increases. However, we must emphasise the fact that this is a random phenomenon which only leads, as regards gene expression, to statistical frequencies of expression in the cells, i.e. it allows variations in expression between the cells which give rise to differentiation between the cell lines but it is not the sole agent of it. The probabilities for gene expression are optimised, but differentiation also involves cell selection which makes the process even more effective.

We have considered the case of a protein regulator. The model functions identically whatever the chemical nature of the regulator element. If it is a DNA sequence situated far away on the chromosome (Fig. 31), it is the random folding of the DNA molecule by diffusion which determines the probabilities of gene activation. Obviously this is an extremely simple example, but it nevertheless reflects the case of the genes for globin. Their expression is regulated by a DNA sequence situated at a distance on the chromosome called the locus control region (LCR). In the course of development, they are sequentially expressed in the same order as their positions on the chromosome, relative to this LCR. This regulation is due to a phenomenon in which the globin genes randomly compete for the LCR (Townes and Behringer, 1990) depending on their position on the chromosome (Hanscombe *et al.*, 1991; Kupiec, 1996).

Of course, there are many more transcription regulators and genes in a cell nucleus than in our example, but the same probabilistic mode of functioning can be applied overall whatever the number of molecules or their chemical nature. From the moment there are fewer transcription regulator molecules than their DNA binding sequences, these regulatory molecules must compete for their DNA binding sequences, leading to numerous

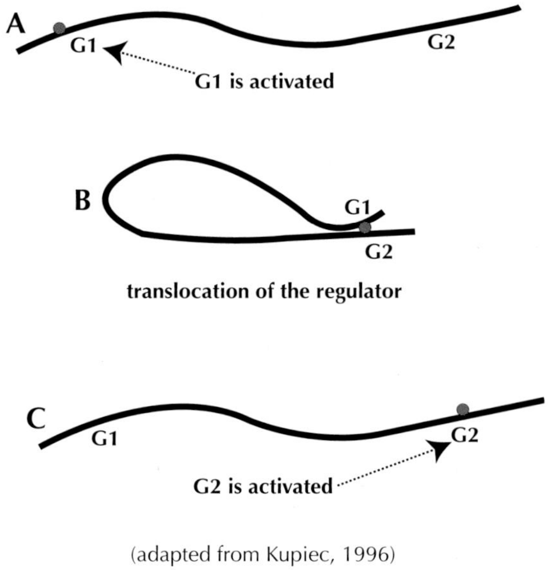

FIGURE 31. Random folding of DNA. DNA movements are themselves subject to thermal agitation. They can therefore induce the translocation of proteins between different genes **(B)** and, thus, their random expression **(C)**. The functioning logic of the stochastic model of expression remains unchanged because the probabilities of interactions between the DNA sites (translocation) depend on the relative positions of the latter.

potential distributions. This condition is readily realised owing to the non-specificity of the interactions between molecules in the chromatin. We have already mentioned the examples of the protein MeCP2 and the transcription factors encoded by homeogenes (chapter 4, §4.1.4).

Moreover, chromatin is organised in a three-dimensional rather than a linear structure as in our example, but that does not alter the model's functioning principle. In a real cell, the position of the genes in the three-dimensional nuclear space will still determine their chronological sequence and the probability of their being expressed in the same way as it would do on a linear chromosome. Although the gene expression probabilities are, in this case, no longer linearly and directly dependent on their position in the DNA, the three-dimensional structure of the chromatin results from DNA folding, the possibilities for which are always constrained by the linear structure of the DNA molecules. In three-dimensional chromatin, the linear distance between the genes therefore remains an important parameter determining the probabilities of the genes being expressed.

It is also interesting to note that this model provides an explanation for the fact that more than 95% of DNA seems to have no function in multicellular organisms. This is made up of sequences which are often interleaved between the genes and do not correspond either to proteins or to regulator elements. This phenomenon concerning the amount of DNA has been called the 'C-value paradox' or more commonly 'junk DNA'. Indeed, in our model all the sequences, even if they do not code for proteins, have a role: they determine the relative position of the genes and their probabilities of being expressed during embryogenesis.

DNA therefore continues to exert an important role in onto-phylogenesis. Its structure is the result of the history of the organism which influences its immediate development, but the way it functions does not come into the domain of the theory of genetic programming. It does not contain a 'plan' of the organism. Nor is it a self-organisation mechanism. Chromatin structuring is not a spontaneous phenomenon: it is constrained by cell selection *via* the signals (or selectors) the cell receives and by the DNA structure which is itself the result of natural selection. Gene expression is therefore both intrinsically probabilistic and subject to natural selection.

To conclude, throughout this chapter we have developed a theory which acknowledges in ontogenesis and phylogenesis a single process. The theory surmounts the contradiction in genetic determinism by integrating the history of the organism in its ontogenesis, and it advances new models of cell differentiation and gene expression. We have also seen that it is compatible with many experimental facts and allows us to make testable predictions. However, its probabilistic nature means it runs counter to very old deterministic traditions deeply rooted in biological thought which are opposed to it being accepted. In the next chapter we shall put ontophylogenesis back in place in the history of biology in order to understand the epistemological obstacles which hinder its acceptance.

7

Biology's Blind Spot

SUMMARY. Heredity and generation have always been dominated by two main ideas. In the Hippocratic conception, the organism reproduces itself in its totality. Particles are released by each of its parts and are rearranged into a new organism, during ontogenesis, by a property of spontaneous organisation, as in holism. This theory, initially put forward by Hippocrates, implies that acquired characteristics are inherited. It survived until the 19th century in various more or less elaborate forms, the last in line being Darwin's. Aristotle criticised Hippocrates' theory because as far as he was concerned, matter would be incapable of organising itself. According to the Aristotelian conception, the organism does not reproduce itself. It is an actualisation of its Form, which also corresponds to its immortal essence. This notion reappeared at the end of the 19th century with genetics. The information that this theory advances as a key concept in its modern version is an update of the Aristotelian Form. The contradiction in genetic determinism was already present from the beginning of genetics, even before molecular biology arrived on the scene. As Morgan explained, it is impossible to establish specific relationships between genes and phenotypic characteristics. Despite this fact, the genetic theory has been staunchly maintained throughout the 20th century, its persistence being explained by extra-scientific reasons, its essentialism guaranteeing humankind its identity and privileged place in nature. Because it is intrinsically a probabilistic process subject to natural selection, ontophylogenesis resorts neither to the essentialism of genetics nor to a Hippocratic type of concept.

Ever since ancient times when living beings began to be studied, heredity and embryogenesis have raised issues and incited uninterrupted debate during which many different concepts have clashed with each other. In this chapter we shall set out the main terms of this debate. We shall see that the way the problem of generation has been repeatedly posed limits the possibilities of analysing it and that consequently, biology is a prisoner of an Aristotelian concept which prevents us grasping ontophylogenesis.

To achieve our end, we must enter the field of metabiology. Indeed, any science is always based on an ontology. Independently of any experiment, it supposes to be real prime entities making up the world. The choice of these initial entities is fundamental because they form the solid basis on which science can develop and condition all the theories that it is later possible to construct. To have an in-depth understanding of the problems of biology that we have mentioned and to be able to overcome them, it is necessary to understand its ontological foundations. In fact, metabiological questions are implicit in all our developments and surface throughout our analysis. We shall now try to tackle them head on.

The history of biology is tortuous. It is full of theories which succeed each other mixing different points of view and progressively evolving with a variety of nuances. We are not attempting here to retrace the complete history of the theories of generation. That would need a much more exhaustive study. However, it is possible to pick out the origin in Antiquity of the problems that we have encountered in this book, in the confrontation between two extreme theories which have since been taken up by numerous authors in just as many variants, sometimes mixing elements of one with those of the other. We shall start by describing the two principal concepts, explaining their antagonistic relationship. The first was expounded by Hippocrates of Cos (460–377 B.C.); it was then refuted by Aristotle (384–322 B.C.) who formulated the second. The importance of the questions raised by these two thinkers and the difficulty of resolving the debate that they initiated is shown by the fact that after having survived in various forms, these questions resurfaced in the 18th and 19th centuries under the respective names of the pangenetic and genetic theories, and again came into conflict, as happened in Antiquity. The pangenetic theory was propagated by, among others, Buffon (1707–1788), Pierre Louis Moreau de Maupertuis (1698–1759) and Darwin. The genetic theory was initiated by Gregor Mendel (1822–1884), August Weismann (1834–1914) and Hugo De Vries (1848–1935) to become the theory dominating the whole of biology in the 20th century, following the work of Thomas Morgan (1866–1945). After recalling the history, we shall return to the current problems to free ourselves from these former ways of thinking and position ontophylogenesis in this debate.

Our analysis will reveal to what extent modern biology is still impregnated with pre-scientific essentialism, hindering its development. This essentialism presents the Form as the prime entity and one that it seems impossible to go beyond, and gives rise to the contradiction in genetic determinism. We shall see that this impasse originates in the belief we have in the reality of the species. We are blinded by what seems absolutely obvious, and this leads us to see the species as the insurmountable horizon of biological thought (Kupiec, 1999, 2004).

7.1 Generation according to Hippocrates

Hippocrates is considered to be the founder of medicine. He lived in Greece in the 5th century B.C. His writings are grouped into a *Hippocratic Collection* consisting of about sixty treatises concerning a great many aspects of medicine and biology, but a large part of these treatises were in fact written by his disciples. Hippocrates' medicine was for a long time very influential. It is based on the physiological theory that there are four humours: phlegm, blood, yellow bile and black bile, and that imbalance between these humours inside the organism causes disease. A major influence is also attributed to the environment.

Two treatises of the *Hippocratic Collection* entitled *On Generation* and *On the Nature of the Child* tackle the questions of heredity and embryogenesis which concern us. According to Hippocrates, the sperm contains humour emanating from all parts of the body. "*Friction on the penis and the movement of the whole man cause the fluid in the body to grow warm: becoming diffuse and agitated by the movement it produces a foam...*" (OG 1.2). This physical emanation passes via the brain and spinal cord to reach the testes. An identical process produces the female seed. "*In the case of women, it is my contention that during intercourse the vagina receives friction and the womb is disturbed, an irritation is set up in the womb which produces pleasure and heat in the rest of the body. A woman also emits something from her body*" (OG 4.1). The mixture of the two seeds in the uterus causes conception, but the quantities of seed coming from the two parents are not identical for all the parts of the body, and quantitative dominance between what has come from each parent determines the characteristics of the new child. Thus to determine the sex: "*...both partners alike contain both male and female sperm... (...)...the resultant sex is determined by whichever sperm prevails in quantity*" (OG 6.1). The same determination principle applies to all parts of the body.

> "*If from any part of the father's body a greater quantity of sperm is derived than from the corresponding part of the mother's body the child will, in part, bear a closer resemblance to its father; and vice versa*" (OG 8.1).

There are two aspects in this Hippocratic concept. The foam is produced by the whole body. It replicates itself as a whole and it is this whole which imprints its order. In this sense, it is a theory which anticipates modern holism. In addition, this idea involves material continuity of the body which persists through reproduction.

Each part replicates itself. The foam forming the seed contains direct excrescences of each organ which are then amplified during embryogenesis retaining the characteristics of the parent from which they come. There is never any material interruption in the succession of organisms. They reproduce directly, one from another, the child from the whole parent (Fig. 32), a concept which has a

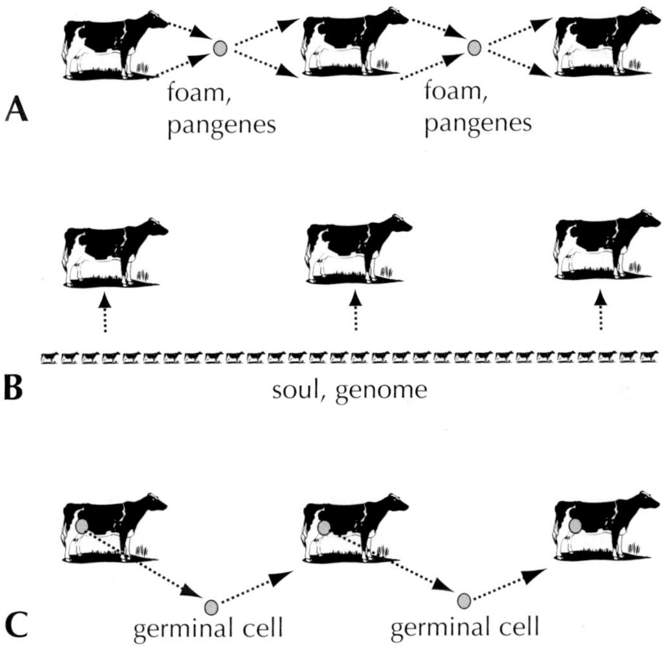

FIGURE 32. **The conceptions of heredity. A:** The Hippocratic conception is holistic and continuist. The entire body is reproduced through foam or pangenes. **B:** The Aristotelian conception is reductionist and discontinuist. The body is a lateral excrescence that is formed from the soul or genome. **C:** Ontophylogenesis does not come into either of these two schemes. The body is formed from a germinal cell. It is a part of it but it does not carry the total representation of it.

very important consequence: it implies that characteristics acquired are inherited. Since the parts of the body can be modified by the way of life of the organism, the foam which comes from these parts might also be supposed to carry these acquired modifications. This leads Hippocrates to envisage the possibility of the organism having a role in hereditary transmission. For example, in the case of children born to deformed parents he writes, "*But when there is some disease involved, and the four innate species of humour from which the seed is derived, form sperm which is not complete, but deficient in the deformed part, it is not in my opinion anomalous that the child should be deformed similarly to the parent*" (OG 11.1).

In the treatise *On the Nature of the Child*, Hippocrates also puts forward a theory of embryogenesis which complements his theory of heredity. In the first phase of embryonic development, purely physical and chemical processes are in action. Breath and heat, as he understands them, play a preponderant role. After coitus therefore, the seed from the two parents mixes "*...and gathers into one mass which condenses as a result of the heat. Next, it acquires breath, since it is in a warm environment*" (ONC 12.1). The seed thus acquires its own ability to respire. It is heated and "*As it inflates, the seed forms a membrane around itself; for its surface, because of its viscosity, stretches around it without a break, in just the same way as a thin membrane is formed on the surface of bread when it is being baked...*" (ONC 12.6). In the second phase, the tissues and organs differentiate from this primitive embryonic structure by a principle of spontaneous organisation. Attraction between like and like draws identical parts together and unites them.

> "*As the flesh grows it is formed into distinct members by breath. Each thing in it goes to its similar — the dense to the dense, the rare to the rare, and the fluid to the fluid. Each settles in its appropriate place, corresponding to the part from which it came and to which it is akin*" (ONC 17.1).

Finally, these theories of reproduction and embryogenesis culminated in an approach to human physiology based on the principle of balance between the four humours. Hippocrates made frequent use of botanical analogies. Each of the four humours has its own source: the head for phlegm, the heart for blood, the gall bladder for bile, and the spleen for water, and in the same way that any plant draws its specific humour from the earth, each of the sources draws its own humour from the stomach. The humours circulate between their reservoirs and the body as needed and equilibrium is established according to the principle of communicating vessels. Disease arises from an imbalance in this relationship.

In this Hippocratic concept, the material continuity of the body is ensured by cyclic processes alternating phases of sublimation of the foam, which represents all parts of the body, with phases of condensation of the flesh. During embryogenesis, organisation of the living organism appears through a principle of order inherent in the matter itself, the attraction of like by like, which makes identical elements assemble spontaneously. Global organisation results from interactions which occur between the parts without there being a plan of the whole organism to guide the process. If this theory is positioned relative to present debates and to other theories from Antiquity, particularly Aristotle's, we can say that it has general characteristics resembling those of a theory of self-organisation.

7.2 Generation according to Aristotle

Aristotle lived in the century following Hippocrates. He is widely known for his logic, metaphysics and physics and his name is immediately associated with these areas. In reality and perhaps above all, he was also a great biologist who continues to have a profound influence. A third of the works attributed to him which have reached us are devoted to biology. It is even probable that it was his finalist biology which influenced his physics and metaphysics, and not the reverse, by suggesting the concept of finality at work in all natural processes. This is the principal aspect of his philosophy. For him, everything exists with an end in view and this final cause is the idea which governs the genesis of all things, whether natural or artefacts. The example of this finalist process, frequently given by Aristotle, is artistic production. The cause of a marble statue is the idea of the statue in the mind of the sculptor, the production of it being the objective of all the processes used by the sculptor.

Three other causes are however also at work. All the while he is sculpting the statue, its form already exists in the mind of the sculptor and serves as a model or plan for him. This is the formal cause which guides or organises the production of the statue. In addition, without matter, it would be impossible, so the marble is the material cause, and the work of the sculptor on this marble is the efficient cause transforming the crude marble into a refined form. There are therefore, in all, four causes which are almost indissociably grouped two by two: the final and formal causes direct the process, the material and efficient causes carry them out in concrete terms. At the heart of this Aristotelian concept is the postulate that matter is not capable of organising itself and to do so it needs a form which is the equivalent of a mould or a template to guide the material processes. The form is also the specific essence of a being in this ontology. It is what it really is without all the little accidents which affect it and which differentiate it from other beings belonging to the same species. For example, if we want to understand what a human really is, he must be defined by the essence common to all individual humans, that is to say by all the specific characteristics corresponding to the human species, leaving aside all the small individual differences (big, small, blond, brown, etc.) which are only accidental. In this hylemorphic ontology, explanation by material and efficient (mechanical) processes only has any sense because they are controlled by the form which is itself subject to the final cause. The latter is the end of the process when the form has been produced. If we were to remain with mechanical causes we would be trapped in an infinite succession of causes and effects. In contrast, Aristotle's world is a world of finished processes and this finiteness allows the origin of things to be understood: the ultimate origin of the statue is also the idea of it in the sculptor's mind. This applies to everything that exists. Nature therefore has a plan.

We shall see how this general conception applies to biology. Aristotle worked on the questions which preoccupy us particularly in his treatise *The Generation of Animals*, in which he completely refuted Hippocrates' theory. After setting out the arguments in his favour, he demolished them one by one. We shall concentrate on the one which is most important for putting his conception into context and considering the problems that we are discussing in this book.

The resemblance between parents and children concerns general characteristics such as height or corpulence, as well as parts of the body such as the head or the feet. Aristotle pointed out that according to Hippocrates and his adherents, "*...if then the coming of the semen from the whole body is cause of the resemblance of the whole, so the parts would be like because it comes from each of the parts*" (GA p. 41). Now, these parts are anomeomere, that is to say they are heterogeneous, composed of different homeomeric (homogeneous) elements which are tissues: flesh, bone, hair, nails, etc. The resemblance of one part such as the head or the feet does not depend so much on the resemblance of these tissues as on the way in which they are arranged one with another. In the resemblance between parents and children an organisational element is involved, which is not material and which cannot be transmitted by a simple material excrescence of such parts of the foam, as postulated by Hippocrates' theory. For the latter, "*...the semen would come rather from the elements than anything else, for how can it come from their composition? Yet without this composition there would be no resemblance*" (GA p. 45). Relative to his historical context, Aristotle adopts a similar point of view to that used twenty-four centuries later by Schrödinger to justify the existence of a genetic code in living systems (chapter 3). Both consider that organisation could not come from the simple set of material processes and that an additional principle of order is necessary. This argument is devastating for Hippocrates' theory and allows Aristotle to put forward his own.

He first of all establishes the nature of sperm. He believes it is a unique digested food residue equivalent to blood, which serves to produce all the parts of the body: "*...it is from the blood, when concocted and somehow divided up, that each part of the body is made...*" (GA pp. 66–67). As we can see, his conception differed considerably from that of Hippocrates, in which there was a plurality of foams from all parts of the body which were supposed to be added together in the sperm, each of which could only recreate the part from which it originated. Aristotle believed that there is but one sperm, the substance of which is homogeneous and which is capable of re-forming the whole body. In addition, there is a qualitative differentiation in the role of the two sexes. We saw that for Hippocrates the contribution from the two sexes was identical and symmetrical, their relationship only being governed by a principle of quantitative dominance. For Aristotle this is no longer the case. Because the female seed is less elaborate, it does not contribute in the same way as the male seed: the latter "*...contributes the principle of movement and the female the material*" (GA p. 86). We thus find here the general principles of his hylemorphic ontology applied to biology. On the one hand, there is the female material intrinsically undetermined and incapable of organising itself, while on the other, the male sperm provides the principle of organisation, the form, which structures it.[43]

The resemblance between generations does not, in this concept, arise from reproduction of the whole material body itself, as Hippocrates believed, but from transmission of the formal cause that Aristotle calls the 'soul' of living beings. There is a break in the material continuity of the body which must be completely reconstructed for each generation by the female seminal matter being given structure by the Form. What is perpetuated is not the body itself, but the Form, which allows reproduction of the same structure and engenders the species by identical individuals succeeding each other. Adult living beings do not directly reproduce one from another (Fig. 32).

There is also a model of ontogenesis corresponding to this theory of generation which postulates a hierarchical organisation of living things with increasing levels of complexity (Fig. 11B). This model is described in another of Aristotle's biological treatises called *Parts of Animals*. In the beginning, there is "*composition out of the Elements*" (PA 646a) which are moistness, dryness, heat and cold. Their combination forms the four basic elements: earth, air, fire and water, that are mixed in turn in various proportions to produce the homeomeric parts such as bone, flesh and other tissues. Association of these tissues finally gives rise to the anomeomere parts such as the head, the hands or the feet. For Aristotle, this ontogenesis is guided by the formal cause which determines the

[43] This sexist theory registers, in addition, in the etymology of the terms that we use, since 'material' and 'maternal' have the same root.

formation of the organism by its parts in a way that today we could describe as programmed.

> *"So the best way of putting the matter would be to say that because the essence of man is what it is, therefore a man has such and such parts, since there cannot be a man without them. (...) Because man is such and such, therefore* ***the process of his formation must of necessity be such and such and take place in such a manner; which is why first this part is formed, then that"***[44] (PA 640b).

Finality is projected via form onto the parts. If this were not the case, there would be a risk of their not being correctly put together to form the organism composed of functional organs.

> *"The Cause which I have just stated as controlling the relation between them is, of course, a Final Cause; but when we go on to inquire in what sense it is 'necessary' that they should be related as they are, it becomes clear that they must of necessity have been thus related to each other from the beginning"* (PA 646b).

While Hippocrates' theory thus has similarities with the theory of self-organisation, Aristotle's is closer to genetic determinism. In both these theories, the central idea is that matter cannot organise itself and must be guided by a principle which represents the whole organism. In Aristotle's theory, as we have just seen, it is the formal cause and in genetic determinism the information contained in the DNA which, via the property of stereospecificity, provides the molecules with order (see chapter 3).[45] In both cases, the organism is constructed owing to this principle of order with a hierarchy of increasingly complex levels.

---

[44] Original text not in bold.

[45] The etymology again supports our analysis. Information means, literally, 'giving form'.

This resemblance between genetic information and the formal cause has already been pointed out by many authors, biologists or philosophers, including emphatically by the founder of molecular biology, Delbrück himself (Delbrück, 1971; Mauron, 2002; Mayr, 1982; Vinci and Robert, 2005), so indeed this is nothing really new. However, this resemblance is generally accepted as being positive. It is interpreted either as a mark of the genius of Aristotle, who might be considered the forerunner of molecular biology, or as a curiosity which could certainly be suitable for historical analysis, but is of no consequence for concrete biological research. We, on the contrary, think it is an important obstacle which is limiting the development of research right up to its experimental aspects. We have seen that the concepts of information and stereospecificity, which form the core of genetic determinism, lead to a contradiction between this theory and experimental research data. In reality, this contradiction goes much further back. Since it has its roots in the underlying metabiology of classical genetics, it was already present in the founding works of this discipline, well before molecular biology arrived on the scene.

7.3 The pangenetic theory

Despite the enormous accumulation of experimental observations and data which have enriched our knowledge of the living world since Antiquity, biological thought has not made much progress but has remained under the influence of the same ways of thinking. Indeed, it is striking to see that both the Hippocratic and Aristotelian conceptions persisted until the 18th and 19th centuries as so-called pangenetic and genetic theories, and how, in the same way that Aristotle refuted Hippocrates to bring his theory to the fore, the birth of genetics coincided with the pangenetic theory being refuted. This theory circulated in different forms, with Buffon and Maupertuis each producing their own version. For each of these authors their principle was analogous to that of Hippocrates' theory, each part of the body emitting particles or supernumerary molecules which migrate to the reproductive organs to form the seminal matter. Darwin was also the author of a pangenetic theory published in 1868 in his book *The Variation of Animals and Plants under Domestication.* This theory did not occupy a central place in his work, as he was not directly interested in questions of heredity and ontogenesis, but was, for him, more a question of filling a gap. The theory of natural selection needed an explanation of the variations on which selection operated, and therefore of reproduction. Darwin's position was not as clear cut as that of the synthetic theory of evolution which has integrated contributions from 20th century genetics, particularly the existence of random mutations. His concept seems implicitly to call on a theory involving random variation which he could never formulate himself due to the stage of development of knowledge in his time. There is evidence for this point in the first sentence of chapter V of *The Origin of Species*, devoted to the laws of variation:

> *"I have hitherto sometimes spoken as if the variations — so common and multiform in organic beings under domestication, and in a lesser degree in those in a state of nature — had been due to chance. This, of course, is a wholly incorrect expression, but it serves to acknowledge plainly our ignorance of the cause of each particular variation"* (OS p. 173).

Darwin did accept as causes of hereditary variation both the direct influence of the environment and the use and disuse of organs (OS pp. 173–204). These factors are usually considered to be typically Lamarckian,[46] including by the most orthodox Darwinians. The classic example given to illustrate the effect of these factors is the neck of the giraffe which elongates as it tries to eat leaves at the top of the tree. The presence of these Lamarckian elements in Darwin has already been underlined by André Pichot (Noble, 2006; Pichot, 1993). However, Darwin also thought that in many cases variability was 'indefinite' and 'fluctuating' (Darwin, 1868). He considered

[46] Incorrectly, as Lamarck, like Darwin, only conformed to the generally accepted opinion on the subject.

that, even if there was a cause, e.g. a change in the living conditions of the organism, the result was not necessarily uniform, variations induced by the same cause possibly differing one from another. In this respect, he anticipated mutation as we understand it today.

Darwin's theory of heredity, which he himself qualified as a provisional hypothesis, is particularly interesting because it shows the persistence of the pangenetic conception in the period that immediately preceded the rapid development of genetics. At that time biology had already made major progress, in particular with the discovery of the cell which was now accepted as the basic unit of living organisms. It is fascinating to see to what extent Darwin's theory still resembled that of Hippocrates, despite these developments. Here in his own words is his main hypothesis.

> *"It is universally admitted that the cells or units of the body increase by self-division or proliferation, retaining the same nature, and that they ultimately become converted into the various tissues and substances of the body. But besides this means of increase I assume that the units throw off minute granules which are dispersed throughout the whole system; that these, when supplied with proper nutriment, multiply by self-division, and are ultimately developed into units like those from which they were originally derived. These granules may be called gemmules. They are collected from all parts of the system to constitute the sexual elements, and their development in the next generation forms a new being; but they are likewise capable of transmission in a dormant state to future generations and may then be developed"* (Darwin, 1868, Vol. II, pp. 369–370).

The same principle is still of emanations from each part of the body allowing its global reproduction (Fig. 32), but Darwin's theory is nevertheless more sophisticated on certain points. It supposes that the gemmules are secreted not only in the adult stage but at all stages of development. These unite with the cells to give them their specific character and embryogenesis is thus guided by these minuscule granules corresponding to each stage. Darwin was aware of the cell theory which he tried to integrate into his conception of heredity, but from a functional point of view, the gemmules are still analogous to Hippocrates' foam.

While being relatively marginal in Darwin's work, the Hippocratic conception of heredity is nevertheless not fortuitous. It would have been difficult for Darwin to uphold an Aristotelian theory: as Jean Gayon (1992A, 1992B) saw it, to be able to conceive of its transformation, he had abandoned the idea that the species corresponded to a structure or form shared by a population of individuals. He had even completely refuted this concept. In his opinion, all the individuals of a species differ from each other and it is these individual differences which provide the ground on which natural selection acts.

> "*...we have many slight differences which may be called individual differences, such as are known frequently to appear in the offspring from the same parents, or which may be presumed to have thus arisen, from being frequently observed in the individuals of the same species inhabiting the same confined locality.* ***No one supposes that all the individuals of the same species are cast in the very same mould.***[47] *These individual differences are highly important for us, as they afford materials for natural selection to accumulate, in the same manner as man can accumulate in any given direction individual differences in his domesticated productions*" (OS pp. 101–102).

It is therefore difficult for Darwin's theory to adapt to the Aristotelian conception. The latter supposes that individuals of a given species are born due to transmission of a form which by definition does not vary. If this were the case, how could they evolve and give rise to a new species? For this reason, Darwin demonstrates moreover that all the characteristics of an organism, including those which are considered to be important from a

[47] Original text not in bold.

systematic or functional point of view, can give rise to individual variations.

> *"These individual differences generally affect what naturalists consider unimportant parts; but I could show by a long catalogue of facts, that parts which must be called important, whether viewed under a physiological or classificatory point of view, sometimes vary in the individuals of the same species. I am convinced that the most experienced naturalist would be surprised at the number of the cases of variability, even in important parts of structure, which he could collect on good authority, as I have collected, during a course of years. It should be remembered that systematists are far from being pleased at finding variability in important characters..."* (OS p. 102).

Darwin went a long way with his analysis as he even came to doubt the objective reality of species. He suggested that their classification only depended on the subjectivity of the classifier.

> *"From these remarks it will be seen that I look at the term species as one arbitrarily given, for the sake of convenience, to a set of individuals closely resembling each other, and that it does not essentially differ from the term variety, which is given to less distinct and more fluctuating forms. The term variety, again, in comparison with mere individual differences, is also applied arbitrarily, for convenience' sake"* (OS p. 108).

This nominalist point of view is easily understood in the context of his theory. For Darwin, the species is the result of a process of natural selection which only serves to amplify individual differences. Resemblance between individuals is an indicator of genealogical proximity: classification can only be differential. The concept of species picks out the differences between populations of organisms which appear after their multiplication and not an essential identity of these populations based on the transmission of an unvarying structure. The species must therefore be understood as a genealogical community.

"In this chapter I have attempted to show that the subordination of group to group in all organisms throughout all time, that the nature of the relationship, by which all living and extinct beings are united by complex, radiating, and circuitous lines of affinities into one grand system ... all naturally follow on the view of the common parentage of those forms which are considered by naturalists as allied, together with their modification through natural selection, with its contingencies of extinction and divergence of character. In considering this view of classification, it should be borne in mind that the element of descent has been universally used in ranking together the sexes, ages, and acknowledged varieties of the same species, however different they may be in structure. If we extend the use of this element of descent — the only certainly known cause of similarity in organic beings — ***we shall understand what is meant by the natural system: it is genealogical in its attempted arrangement, with the grades of acquired difference marked by the terms varieties, species, genera, families, orders, and classes****"*[48] (OS pp. 432–433).

As we can see, Darwin's thinking is very far from Aristotle's. On the other hand, before analysing genetics, we must emphasise the extent to which his ideas approach those of Bernard. These two authors lived at the same time and they both left decisive marks on their discipline, thrusting it into modernity. It is not usual to consider them together because they concerned themselves with very different subjects, yet they have in common a totally anti-essentialist vision. We have just discussed Darwin's position in regard to the species and we have seen that, for his part, Bernard questioned the objective reality of functions (chapter 6, §6.1.3). His anti-essentialism led him to formulate a more radical position, since he put the very notion of the individual organism into perspective in order to enhance the idea of a biological continuum very close to the notion of the genealogical line in his analysis of morphogenesis:

[48] Original text not in bold.

> *"Thus all morphological development is contained in the previous state. This work is pure repetition; it does not have its reason at each instant in a force currently active; it has its reasons in an anterior force. There is no morphology without predecessors.*
>
> *In reality we do not witness the birth of a new being; we see only a periodic continuation. The reason for this apparent creation is therefore not in the present; it is in the past, at the beginning. We cannot find it among the secondary or actual cause; it must be sought in the primary cause.*
>
> *The living being is like the planet that describes its elliptical orbit in virtue of an initial impetus..."* (LPL pp. 240–241).

A little farther on, in one of his rare allusions to the question of the species, Bernard adds:

> *"It is unnecessary to see in this tendency to return to the starting point any particular mysterious force that watches over the conservation of the species. If things happen in this way it is because the being is in some way imprisoned by a series of conditions which it cannot escape, since they are always repeated in the same way outside and inside it"* (LPL pp. 241–242).

The birth of modern evolutionary biology and of modern biology of the organism is thus linked, in the 19th century, to an anti-essentialist point of view which rejects the hylemorphic ontology. As first principle, this point of view substitutes the idea of the genealogical line or biological continuum for the notions of the individual organism and the species based on the concept of Form. Nevertheless, Form would very rapidly return with a vengeance with the advent of genetics.

7.4 The return of Form

Quite a few researchers have contributed to the rapid development of genetics, but Weismann occupies a dominant position. He introduced fundamental concepts which marked a radical break with the pangenetic theory. As Aristotle had done with Hippocrates' theory, in his *Essay on heredity and kindred biological problems* (1891) Weismann dismantled the theory of gemmules and the inheritance of acquired characteristics, then put forward his own theory. It postulates the complete separation of somatic cells (the body) and germinal cells (reproductive gametes), the latter supposed to be the only ones responsible for heredity.

> *"In these animals the power of reproduction is connected with certain cells which, as germ cells, may be contrasted with those which form the rest of the body; for the former have a totally different role to play; they are without significance for the life of the individual, and yet they alone possess the power of preserving the species. Each of them can, under certain conditions, develop into a complete organism of the same species as the parent, with every individual peculiarity of the latter reproduced more or less completely"* (EH p. 73).

More precisely, for Weismann, heredity is due to the transmission of a particular molecular structure contained in the germinal cells, which he called the 'germ-plasm' and which prefigures our DNA. "*I propose to call it the theory of 'The Continuity of the Germ-plasm,' for it is founded upon the idea that heredity is brought about by the transference from one generation to another, of a substance with a definite chemical and above all molecular constitution*" (EH p. 170).

There are several important points in this founding hypothesis which condition the entire coherence of genetics, including its current developments. The first is the separation of the 'soma' and the 'germ-plasm' which prevents reproduction being influenced by the organismic context or the environment. This separation has been continued into modern genetic theory with the two concepts of the phenotype and genotype: the phenotype is the actual organism, corresponding to a set of characteristics controlled by the genotype, which corresponds to the set of genes. Reproduction of the phenotype-body does not occur from the organs as in Hippocratic theory, but through the intermediary of the genotype carried on the chromosomes. Strict determinism ensues in the relationship which links the phenotype to the genotype, which was already present in Weismann's work. "*From the moment when the phenomena which precede segmentation commence in the egg, the exact kind of organism which will be developed is already determined — whether it will be larger or smaller, more like its father or its mother...*" (EH p. 104). This determinism is absolutely necessary for the reproduction of the phenotype from the genotype (or the body from the germ-plasm). It has been regularly restated throughout the history of genetics, right up to the 1960s, with molecular biology and its 'central dogma' stating that the organism is entirely coded by information contained in the DNA (see chapter 3).

A second element in the genetic postulate has more important consequences: the unvarying transmission of the germ-plasm (or DNA in current genetics) ensures continuity of the species through juxtaposing identical ontogenesis, each individual produced being a lateral excrescence of the germ-plasm (Fig. 32). As Weismann put it:

> "*...in each ontogeny, a part of the specific germ-plasm contained in the parent egg-cell is not used up in the construction of the body of the offspring, but is reserved unchanged for the formation of the germ-cells of the following generation*" (EH p. 170).

These germ cells "*only contain the undying part of the organism — the germ-plasm*" (EH p. 209), that is to say, its Form, or soul, according to Aristotle's terminology. The reproductive material is protected, in this conception, from the world and accidents. When Weismann located it in the nucleus of germinal cells, he talked of it as the essence of the cell.

Genetics thus re-established an Aristotelian theory. The body no longer reproduces directly from its parts, but from a germ-plasm of unique origin, sheltered from any fluctuation in the nucleus of the germinal cells. It represents the whole organism and guides its ontogenesis in each generation. It acts as the formal cause of embryogenesis to give the living organism its specific organisation. It is the equivalent of the soul (Form), of what today we call the genetic programme, genetic information or the plan of the organism. The immense accumulation of experimental knowledge which has occurred since the 19th century has provoked major changes in genetics. The genome has replaced the germ-plasm, and the field has incorporated all the discoveries made in the 20th century, though its hylemorphic metabiology has remained the same. The idea that hereditary material contains in-**form**-ation, a true sanctuary preserving the integrity of the species, is at the heart of the concepts of molecular biology. Jacques Monod expressed it in words very close to those of Weismann, encompassing the concepts and language of information theory. According to him, reproductive invariance is the first principle of life which he defines as:

> "*...their ability to reproduce and to transmit* ne varietur *the information corresponding to their own structure; very valuable information, since it describes an organisational scheme which is exceedingly complex and also preserved intact from one generation to the next. (...) the 'invariance content' of a given species is equal to the amount of information which, transmitted from one generation to the next, ensures the preservation of the specific structural standard*" (CN pp. 23–24).

Genetics certainly accepts variation, but as a mutation of this unvarying Form. As Monod again explains, mutation is an accident which upsets reproductive invariance:

> *"Nor, without violating the laws of physics, could the mechanism of replication be completely immune to disturbances, or accidents. (...) We say that these events are accidental, due to chance"* (CN pp. 109–110).

7.5 The contradiction in genetic determinism is a consequence of genetic essentialism

Original Darwinism and genetics are based on clearly antithetical metabiologies. Darwinism is a theory privileging the point of view of evolution and variation, while genetics is a theory which privileges the point of view of ontogenesis and the invariant form. For this reason, synthesising the two theories was very difficult and did not occur spontaneously. It only became possible in the middle of the 20th century and its story has already occasioned in-depth studies which highlight the difficulties encountered (Gayon, 1992A, 1992B). What is of particular interest to us here is the question of the coherence between Darwinism and genetic determinism, and from this point of view, the price paid for synthesis was very high. To preserve a minimum of coherence, DNA has had to be attributed a role of omnipotent governor of biological processes which brings into play very different mechanisms according to circumstances. It allows evolution, through its random mutations, on the basis of a probabilistic mechanism, and also directs ontogenesis by functioning as a deterministic programme (a Form) which eliminates molecular chance.

We have seen the consequences of reintroducing an Aristotelian conception. For DNA to be able to play its role as genetic information, ontological separation between the physical/chemical and biological processes has to be acknowledged. While the first are subject to a principle of order from disorder, the second are subject to one of order from order. In addition, to give an effective content to this principle of order from order, molecular biologists have been obliged to postulate the stereospecificity of interactions between biological molecules, from which arises the contradiction of genetic determinism.

This persistence of hylemorphic ontology, despite the problems that it raises, leads to a question. Does it express, as is generally accepted, the relevance of this ontology for treating living organisms, or is it not rather a symptom of our inability to formulate an appropriate theory? Indeed, not only has hylemorphic ontology repeatedly reappeared in the history of biology, whereas it was abandoned in physics at the time of the Copernican revolution, but it has also been maintained in the 20th century in the form of the genetic theory, despite this theory having been contradicted from the beginning by experimental data.

The contradiction in genetic determinism was, in fact, evident in the earliest work by classical geneticists.[49] The evolution which Morgan's work underwent illustrates this in a remarkable way: he was obliged to considerably modify his initial theory because of the results of his own research. First of all, he had reaffirmed a form of genetic determinism identical to Weismann's. His position was very radical. His belief in the causal ability of genes led him to put the importance of embryonic development into perspective. He thought that the relationship between the genes and phenotypic characteristics was completely determined; knowledge of this relationship was enough in itself to manipulate phenotypic characteristics, without knowing the mechanisms of embryonic development. In his book *The Theory of the Gene* (1926), he asserted:

> *"The theory states that the characters of the individual are referable to paired elements (genes) in the germinal material that are held together in a definite number of linkage groups... (...) The theory of the gene, as here formulated, states nothing with respect to the way in which the genes are connected with the end-product or character. The absence of information relating to this interval does not mean that the process of embryonic development is not of interest for genetics. A knowledge of the way the genes produce their effects on the developing individual would, no doubt, greatly broaden our ideas relating to heredity, and probably make clearer many phenomena that are obscure at present... (...) There is, nevertheless a fundamental assumption implied in the statement just made, viz., that the development follows strictly causal laws. A change in a gene produces definite effects on the developmental process. It affects one or more of the characters that appear at some later stage in the individual.* ***In this sense, the theory of the gene is justified without attempting to explain the***

[49] Before molecular biology, classical geneticists studied the transmission of characteristics independent of the underlying mechanisms.

> ***nature of the causal process that connects the gene and the characters"***[50] (Morgan, 1926, pp. 25–27).

However, Morgan very quickly realised that this determinism was not compatible with experimental reality. Several phenomena blurred the correspondence and causal relationship between the gene and the characteristic. In pleiotropy, a gene affects several different phenotypic characteristics, while in polygeny, it is the reverse, with several genes affecting a single characteristic. There is also the conditional expression of a gene. A characteristic depends on a gene, but it is only expressed under certain environmental conditions, e.g. at a given temperature. More recently variable expressivity has been shown. A gene corresponds to several phenotypic characteristics which are expressed with a certain frequency in a population of organisms. Because of these phenomena, the simple correspondence between genes and characteristics that Morgan talked of in *The Theory of the Gene* is very difficult, if not impossible, to establish. However, this is one of the pillars of genetics. If it is not possible to establish the map of causal relationships between genes and characteristics, genetics is invalid as an explanatory theory. For it to remain significant, the relationship between the gene and the phenotypic characteristic cannot be reduced to a simple statistical correlation seen empirically. This led Morgan to write another book, *Embryology and Genetics* (1934), in which he tackled this question and re-evaluated the importance of embryonic development in heredity. Here is what he said in 1934, less than 10 years after publishing *The Theory of the Gene*:

> *"In the early days of genetics, i.e., at the beginning of the century, 'unit characters' were supposed to furnish the basis for genetic work, and by inference each gene was supposed to produce a specific effect in only one character at a time. This premature inference was very soon found to be erroneous when the manifold effects of each genic change came to be known. It is true that in most*

[50] Original text not in bold.

> *genetic work one particular character is selected as the symbol of the gene concerned with its appearance, but this selection is only because the character selected is the most easily identified, or one that is less variable, i.e. less affected by environment. The next point that calls for consideration is that each character of the adult is the product of many genes,* ***or it may even be said of all the genes if the whole history of the affected organ is retraced to the egg***"[51] (Morgan, 1934, pp. 16–17).

This is a complete change of perspective compared with *The Theory of the Gene*: Morgan goes a very long way in re-evaluating the relationship between the gene and the characteristic. If, as he says, a characteristic indeed depends on all the genes, the basic postulate of genetics — the idea that a phenotypic characteristic can be associated with a gene (or with a limited number of genes) — collapses, because if we extend the observation to its limit, it means that all characteristics depend on all the genes! Certainly, Morgan did not go that far, and to cope with this theoretical difficulty he introduced the idea of genic balance: "*The central idea of genic balance is that all genes are acting, and what is produced is the sum total of their influence*" (Morgan, 1934, p. 17). Since Morgan, systematic analysis of the determinism of genes has only confirmed his interpretation.[52] All geneticists know that cases of an unequivocal relationship between a gene and a characteristic similar to those described by Mendel are very rare exceptions, if they exist at all. However, most of them nevertheless do not question genetic theory. They overcome this difficulty by adding hypotheses which are supposed to complete it. In the wake of Morgan, they imagine that genes act in combinations and in addition they acknowledge the influence of the environment which can affect genetic determinism. Several possible phenotypes correspond to a given genotype, each occurring in a particular environment (Dobzhansky, 1970, pp. 33–36).

---

[51] Original text not in bold.

[52] See, for example, all the genes implicated in so-called genetic diseases isolated in the recent past (Wolf, 1997).

In general, all the elements of ontogenesis which are added to genetic determinism are called 'epigenetic factors', some authors going so far as to reintroduce Lamarckian type mechanisms (Jablonka and Lamb, 1995).

There is a question that must be asked here. Is a simple rearrangement of the genetic theory enough to solve the problem which confronts it, or is a complete change of conceptual framework necessary? Indeed, what is being questioned is the very heart of this theory. The problem raised by Morgan is that of the specificity of the relationship between genes and the characteristics associated with them. If this relationship is not specific, we cannot assert that genes determine characteristics. They may certainly form an important part of the process of embryogenesis which builds these characteristics, but they are not the cause in the sense implied by genetic determinism. Epigenetic mechanisms being added to the action of genes changes nothing regarding this fact (Kupiec, 2001).

This problem is exactly the same as the problem of the stereospecificity of molecules. For the relationship between a gene and its phenotypic characteristic to be specific, the underlying molecular mechanisms must be specific too. However, we have seen that analysis of the action of genes at molecular level, far from resolving the problem, only amplifies it. We must then ask ourselves why, when the absence of specificity in the way the gene acts has already been demonstrated, has this postulate not only been maintained, but has even been extended to the molecular level? Why, too, when the non-specificity of molecular interactions has been demonstrated time and time again, do biologists transfer the property of specificity to the macroscopic level without concern for the theoretical incoherence that this represents (see chapter 4), instead of changing the theory? According to the usual criteria of scientific practice, a theory which has been invalidated in such a radical manner should be abandoned.

There is something very important for biology in regard to this question which differentiates it from physics and explains its difficulty in going beyond hylemorphic ontology. The underlying problem concerns the reality of species. Hylemorphic ontology asserts that the species is a real structure which acts as a cause (the form or the genotype, in the case of genetics). It supposes that the world is intrinsically ordered. Under the apparent diversity of existing things are hidden forms on which they depend. In molecular biology, this is expressed by the principle of order from order advanced by Schrödinger. It comes into play through the genetic programme which corresponds to the plan of characteristic organisation of the species (to its Form). In physics, hylemorphic ontology has been abandoned. Species are considered to be the result of processes and not their cause. They do not correspond to an order given in advance, intrinsic to the world. This is what Schrödinger expressed with the principle of order from disorder. In biology, hylemorphic ontology which makes specificity a central concept is still at work, even though it is the antithesis of Darwinism and despite the problems that it induces. Where does this persistence, this determination to retain it against the experimental evidence, we might say, come from?

7.6 Beyond the species

On taking stock we can see that with the species we have a problem which goes beyond scientific rationality. The nature and objective reality of biological species seem indisputable. They appear to us to be absolutely evident and impossible to question. Are we not being blinded here, however, by our narcissism and our egocentricity? When we deal with species we are also dealing with the human species, therefore ourselves. To abandon essentialism and hylemorphic ontology is to deny the objective reality of the species, which may undermine our own image and the position that we attribute to ourselves among the entities which people the world.

In truth, essentialism reassures us. It tells us that there is meaning to our existence, that there is a nature to which we conform (our essence, our genetic code) and that we have by right a place in the universe (in general, at the summit of a stratified model of the world). Denying the species amounts to denying the idea of this human nature and destroying the foundations of our identity. We will no longer be 'at home in the universe' as Kauffman said (1995) but would be placed in a radically strange situation, returned to the same rank as other beings, including inanimate objects. This is very difficult for us to bear, even unthinkable, and results in the mental block which makes it difficult to abandon hylemorphic ontology in biology. Yet if our objective is to construct a rational biological theory, we must analyse this question more rigorously and avoid being dominated by our subjective and psychological feelings.

Biology's current ontology is based on the concept of specificity. We must go beyond this naïve point of view and build 'aspecific' biology. Does that seem absurd? There are nevertheless a great many examples of scientific theories built on counterintuitive propositions.[53] It is not a question of denying the reality of the species to remain in an academic quarrel, but of putting forward a theory which avoids using a principle of order analogous to the formal cause. Furthermore, Darwin and Bernard have already shown us the way here. The former replaced the essentialist definition of the species by a genealogical definition and the latter discarded finalism in the notion of physiological function.

On the other hand, returning to a Hippocratic type of theory is not a valid option. Such theories make the organism the origin of ontogenesis as it emits the foam[54] from its different parts. This foam is the reflection of the pre-existing organised structure (the organism). Like Aristotelian theories, they thus depend on a principle of order from order. They must always rely on an organising principle intrinsic to matter, similar to holistic principles (attraction of like by like in Hippocrates, Elsasser's holistic memory, emergent properties or the spontaneous tendency towards self-organisation (see chapter 5).

In these respects, ontophylogenesis goes beyond this blind spot of biology without regressing towards prescientific points of view.

[53] It has been a distinctive feature of a scientific theory probably since Copernicus.

[54] Or its equivalent in the different variants of the Hippocratic conception.

Because it unifies ontogenesis and phylogenesis, it is neither Hippocratic nor Aristotelian in concept. Here, the organism is constructed from the global structure of the germinal cell including the DNA and protein complexes and there is therefore material continuity. The germinal cell belongs to the two organisms, the old one which is being reproduced and the new one which is being formed, but it is not the reflection of the totality of the organism. It does not receive molecules from all parts of the body, as is the case in a Hippocratic theory; it is the result of its own history as a cell within the organism. Each ontogenesis is therefore also the excrescence of the germinal line as in Aristotelian theory (Fig. 32).

Ontophylogenesis allows us to escape from the fetters created by these two types of theory in which biological thought has been trapped throughout its history; and if it provides this new perspective, it is because it totally renounces specificity to make room for probability. It does not depend on any principle of order which may be inherent in matter or given *a priori*. The organism is produced in its context by a non-finalist process in which environmental constraints act on intrinsically probabilistic molecular and cellular mechanisms.

It thus forms a radical break because it is based on a new metabiology. While in the Hippocratic and Aristotelian theories the organism is a first principle, really or virtually the subject *a priori* of generation, in ontophylogenesis it is the random result of a process without finality.

Conclusion: A Research Programme and Ethical Principle based on Ontophylogenesis

We have analysed the main points of genetic determinism and seen that it is controlled by what Schrödinger called the principle of order from order. Order is supposed to be inherent in the living organism and expressed in the relationships between the molecules. This order is said to be encoded in the genetic information of the DNA and brought into play by the property of molecular stereospecificity. According to this conception, the proteins unequivocally interact with each other, this interaction being determined by their three-dimensional structure. They fit together like pieces of a puzzle to create the organism, with nothing left to chance. This stereospecificity is necessary for the determinism of the genes to be expressed at macroscopic level in the phenotypic characteristics.

We have seen that throughout the second half of the 20th century, the research programme undertaken by molecular biologists resulted in a very large number of genes and proteins being isolated that are involved in numerous normal or pathological cellular processes, such as gene expression, signalling and cancer. The interaction properties of these proteins with other molecular partners were analysed, and it then appeared that, contrary to the predictions made, these proteins are not stereospecific. More recently, global networks of molecular interactions have been analysed for several species, the results of these studies confirming the importance of molecular non-specificity. Many proteins can interact with more than a hundred partners, which implies that all biological regulation pathways are interconnected and that the way the cell functions cannot be explained just by the structure of the molecular networks, as supposed by the principle of order from order. The networks must themselves be regulated to avoid generalised interference between all cell functions. To resolve this problem, molecular biologists hold that the structure of the cell sorts the non-specific molecular interactions to avoid those which might be harmful and only let those that are really specific occur. However, this means reversing the causal explanation and reintroducing holism. Seen in this light, it is no longer the molecules that determine the phenotype but the reverse: the phenotype of the cell determines the molecular interactions which take place in it. Molecular biology research thus ends in the contradiction of genetic determinism. It denies the theoretical principles which have motivated it and calls for a new conceptual context that integrates these results.

Such developments are perfectly normal in modern scientific practice. Biology needs a theory which grants cell structure a causal role. *A priori*, it could be based on holism. However, the analysis we have carried out has shown us that that reintroduces the idea of animate matter, breaching the principle of the inertia and objectivity of nature on which scientific method is based. Holism also supposes that there are emergent properties implying irrational creation *ex nihilo*, and thus reintroduces a form of hidden mysticism. In addition, despite their opposition on the surface, holism and genetic determinism have a common basis. Both concepts believe that order is real and underlies all phenomena. They believe in the hierarchical organisation of the world supposedly created by superimposed and increasingly complex levels, from which the diversity of things unfurls. The difference between the two stems from the origin of the order. For genetic determinism it comes from below (from the molecules); for holism, order comes from above (from the whole). In the second part of the 20th century, theories of self-organisation tried to give more precise content to holism by suggesting models for application to biology, but they all introduce a new contradiction. In seeking to explain concrete phenomena, they reintroduce, without acknowledging them, external constraints which are applied to systems to give them order, while at the same time proclaiming that these systems are spontaneously organised from the elements composing them. This demonstrates that real organisation phenomena are not self-organisation but hetero-organisation phenomena. Holism and self-organisation are not valid, alternative, theoretical contexts to replace genetic determinism.

Ontophylogenesis resolves the contradictions in genetic determinism and holism because it fully acknowledges the non-specificity of biological molecules and the intrinsically stochastic character which that imposes on the living organism. It substitutes a historical explanation for the explanation involving levels of organisation. In this concept, it is neither the molecular that determines the macroscopic, nor the reverse. Ontogenesis is an extension of natural selection in the internal environment of organisms. A living being is the product of a history in which selective constraints and stochastic molecular events have been integrated into a single process.

We have seen that, based on this conception, more accurate models of cell differentiation and gene expression can be constructed, models which incorporate experimental facts and lead to predictions that can be tested. Computer simulation also demonstrates that this is a relevant general theoretical framework for explaining the principal properties of embryogenesis. However, it is obvious that this concept must not be considered to have produced a perfect theory, and the conclusion of this book therefore opens up two main aspects that we can immediately see should be developed in the future.

The first consists of constructing a much wider research programme, which will certainly be modified as the first experiments are performed. We can nevertheless outline the strategic direction. The variability of biological parameters is a general, undisputable phenomenon that all experimental biologists acknowledge. However, this variability is not recognised by itself as a biological parameter which can have a causal role. In the action of cellular mechanisms, which have always been considered as fundamentally deterministic, it is seen as a margin of fluctuation. Present research programmes, influenced by genetics and molecular biology, are seeking to analyse genetic information and the networks of molecular interactions which ensue from it. This strategy corresponds to having a fixed view of the living organism that ignores variability, and leads inexorably to the contradiction in genetic determinism. We are drawing up maps. After mapping the genome, we are now drawing up one for all the genes transcribed in a cell (transcriptomes) and another for all the proteins with their interactions (proteomes). It is hoped that from this data, we will be able to explain, possibly with the help of a computer program, how the cell functions. This is an error. Accumulating these data is certainly not totally devoid of interest, but the genes expressed and the interactions which are produced between proteins in a cell are the result of the way it functions and not the cause. The interactions have been selected by cellular processes from among the huge number of combination possibilities arising from molecular non-specificity, and it is precisely this selection process, which is the functioning process of the cell, that we need to explain. We come back here to our metaphor of the man lost in the Amazonian forest. The selection process is the Amazon. We have to turn back to see it and analyse it instead of continuing to accumulate contingent observations.

Inevitably, ontophylogenesis radically modifies the approach to mapping the living organism which genetics induces, by giving variability its rightful place. Since it acknowledges the intrinsically probabilistic character of biological phenomena, it equally acknowledges the variability which arises from it and provides the substrate for phenomena of cellular selection. For ontophylogenesis, variability can no longer be reduced to a simple margin of fluctuation. On the contrary, it attributes to it a primordial causal role. Studying it must therefore be systematised and put back into probabilistic explanatory schemes, and this should be done directly in regard to experimental measures set up and not just at the level of theoretical interpretation. Indeed, if variability is a significant biological parameter, it must vary quantitatively during physiological processes, like any parameter, and this quantitative variation must be correlated with other parameters of these processes in such a way that this correlation helps to explain them. The research programme to be set up should therefore aim to raise biological variability from the status of simple fluctuation to the status of a fundamental parameter. This will require its systematic, quantitative study.

The second and final point concerns the possibility of constructing an ethical principle. We have demolished the idea of order intrinsic to the living organism. We have rejected any form of animism. In doing so, we have just returned to the general foundations of scientific method. However, in these conditions, on what basis can we build an ethical principle if there is no natural order in the world to refer to? What principle can we rely on if "Pure chance, absolutely free but blind" as Monod said (CN, p. 110), is the ultimate reality hidden in the deepest depths of ourselves? Ontophylogenesis could be accused of nihilism: demolishing the role of Form leads to doubting human specificity, and ends in radical anti-humanism. Yet this is inexact. Ontophylogenesis does not reject the idea of order in itself, but the idea of an absolute order, transcendent and unalterable. Order exists, but it is relative and can change. It depends on the relationship of the living organism to its environment. The organism can only exist in and through this relationship which it just interiorises in its internal environment, and which spares it precisely the void of absolute chance. The consequence of this is of the greatest importance: the living organism thus comes into being relative to what it is not. That otherness is present, inseparable, in its identity.

There seems to us to be no nihilism here; on the contrary we see the possibility of finding an ethical principle, without resorting to transcendence.

Glossary

Allele: Organisms often possess several versions of the same gene. Each version is an allele. This is particularly the case in diploid organisms in which there are two versions of the same gene each carried by each homologous chromosome (see *Diploid*).

Antibody: See *Immunology*.

Antigen: See *Immunology*.

Aristotelism (the four causes): According to Aristotle, there are four kinds of cause in Nature: the 'material cause' provides the matter out of which a phenomenon is constituted or involved in the ontogenesis of objects. This material has no form on its own; it is incapable of creating anything of an ordered nature by itself. It therefore has to be associated with the 'formal cause' which provides a form corresponding to the essence of the phenomenon or thing. These two causes are however insufficient because there is an absence of movement. The 'efficient cause' is the immediate motor that produces phenomena. The 'final cause' is the underlying purpose of any process, the finality of which is the production of the essence or form of things.

Blastocoel: See *Blastula*.

Blastula: Early stage in the development of embryos characterised by the appearance of a cavity known as the blastocoel.

Brownian motion: See *Thermal agitation.*

Cell culture: Technique consisting of culturing cells artificially in dishes by adding a nutrient medium in sterile controlled conditions.

Cell differentiation: Multicellular organisms possess different types of cells with different structures and functions (muscle, bone, blood, skin cells etc.). These differentiated cells appear during development of the embryo through progressive specialisation from division of the initial egg. In the context of molecular biology, these differentiations are considered to depend on the activity of different genes in each cell type, corresponding to the expression of different proteins. All the cells have the same genome but do not express the same genes.

Cell interaction: Cells influence the activity of their neighbours through their own activity, the molecules they release and the functions they perform.

Chromatin: Entity formed by DNA molecules and the proteins with which they interact (including the histones of nucleosomes. See *Nucleosome*).

Chromosome: Form of chromatin compacted at the time of cell division (see *Chromatin*).

Competition: See *Molecular interaction.*

Computer simulation: Simulation of a phenomenon by a computer with a program which (in part) reproduces the phenomenon from a model, using mathematical (for digital simulations) and/or data processing (for computer simulations) methods. The model can then be studied, varying the parameters at will. The model may be a theoretical representation or represent a mechanism. Its function may also be just to replicate the observable evolution of a system, without worrying about replicating a mechanism or theoretical principle.

Cultured cells: See *Cell culture.*

Cytokines: Proteins with various functions secreted by the cells of the immune system.

Dendrite: See *Neuron.*

Differentiated cell: See *Cell differentiation.*

Diffusion: See *Thermal agitation.*

Diploid: Diploid organisms have two sets of homologous chromosomes, each arising from one of the two parents.

Epigenetic modifications: See *Protein synthesis.*

Essentialism: Philosophy which gives pre-eminence to essences. Generally, these essences are Aristotelian forms. There is said to be immanent order and finality in the world because every thing is determined by its underlying essence which it seeks to manifest. Essentialism and the reality of the species go hand in glove since a species is a set of things having the same essence.

Eukaryote: Organism in which the cell or cells contain numerous organelles delimited by membranes, particularly the nucleus which contains the DNA associated with the proteins in the chromatin (See *Chromatin*).

Evolutionary synthesis (or synthetic theory of evolution): Predominant theory of contemporary biology resulting from the synthesis that occurred towards the middle of the 20th century between original Darwinism and the contributions made by other branches of biology including population genetics. In this theory, evolution occurs through the selection of mutations of genes, an idea which did not exist in Darwin's theory as he knew nothing of genetics. However, this synthetic theory encompasses more than what is known as neo-Darwinism, which developed through the merger, initiated by August Weismann, of Darwinism with the work of the first geneticists at the end of the 19th century.

Ex vivo: Experimentation on a living system out of its normal context, e.g. cells isolated from the organism and cultured independently (see *Cell culture*).

Final cause: See *Aristotelism*, *Essentialism.*

Finality: See *Aristotelism*, *Essentialism.*

Fluctuation in molecular concentration: See *Molecular interaction.*

Form: See *Aristotelism*, *Essentialism.*

Formal cause: See *Aristotelism*, *Essentialism.*

Gene: Determinant of the hereditary phenotypic characteristic carried on the chromosomes. Initially geneticists considered the genes subject to absolute determinism. Nowadays, it is more often accepted that they are influenced by environmental factors (see *Phenotype*).

Genome: The total genetic material (DNA) of a cell.

Genotype: All the genes carried on the chromosomes (see *Gene*).

Growth factors: Small proteins initially described for their ability to activate cell proliferation. They exert pleiotropic effects on several cellular processes (differentiation, survival, apoptosis) in different lines.

Haematopoietic cells: Blood cells.

Heat motion: See *Thermal agitation.*

Hylemorphic ontology: Ontology according to which things are the result of two first principles, matter and form (see *Aristotelism*, *Essentialism*).

Immunology: Discipline which studies immunity, i.e. the capacity of an organism to resist and rid itself of an agent which is foreign to it, e.g. a virus. There are two types of immunity. Humoral immunity involves the secretion of antibodies by B lymphocyte cells. These antibodies are proteins (immunoglobulins) which react with a part of the infectious agent (the antigen) by complexing, neutralising and eliminating it. Cell-mediated immunity involves the action of T lymphocyte cells which, owing to the receptors situated in their cell membrane, directly recognise antigens and neutralise infectious agents.

In vitro: Experimentation outside a living system, e.g. a biochemical reaction produced in a test tube.

In vivo: Experimentation performed on a living organism.

Kinase: Post-translational modification enzyme phosphorylating (adding phosphorus to) proteins (see *Protein synthesis*).

Lymphocyte: See *Immunology.*

Macroscopic: Seen with the naked eye. A phenomenon can be analysed macroscopically (at our level of observation) or at the molecular level. For example, the rain which falls from clouds in drops as we see it, is also behaving as a population of molecules of water.

Molecular interaction: A molecular interaction involves physical contact between molecules that can produce a more or less stable bond. The bond may allow complexes of several molecules (macromolecular complexes) to be constructed which are incorporated into the structure of the cells (membranes, chromatin etc.). It can also trigger the biochemical reactions (enzyme/substrate bond) of metabolism or induce regulation (a bond between a gene transcription regulator protein and a DNA sequence). Competition exists between molecules. Let us imagine red, black and white balls moving randomly in a space. Contacts between balls of different colours occur randomly but their frequency relative to the whole set of balls depends on their relative proportions. For example, the more red balls there are, the more contacts there will be involving red balls. Owing to the random character of the movement of the balls the frequency of these contacts is permanently subjected to random fluctuations. The same is true of molecules subjected to thermal agitation.

Morphogen molecules: Morphogens are chemical substances inducing cell differentiation in the embryo. They often form gradients and exert their effects at defined concentrations.

Morphogenetic gradients: See *Morphogen molecules.*

Morphogens: See *Morphogen molecules.*

Multicellular organisms: Organisms having several cells (see *Cell differentiation*).

Natural selection: Evolution was acknowledged by many naturalists from the 18th century onwards. Darwin gave an explanation that has provided the general framework for modern theories of evolution. Organisms are subject to countless variations. Individuals with advantageous variations (access to food, success in sexual reproduction etc.) reproduce more than others and after several generations become the dominant type. In this way populations of individuals forming species change characteristics and evolve through accumulating variations.

Neo-Darwinism: See *Evolutionary synthesis.*

Neural crest: Region (dorsal to the neural tube) in vertebrate embryos. Its cells migrate in the embryo and give rise to several cell lines including the cells of the nervous system.

Neuron circuits (networks): See *Neuron.*

Neuron: Neurons are cells of the nervous system by means of which nerve impulses are transmitted. In general they are elongated in shape, with many branches (the dendrites) at one end and the other end less branched (ending in synapses). Nerve impulses are transmitted from one neuron to another through the connections which are established between synapses and dendrites. The neurons thus form circuits in which impulses circulate. Owing to the very large number of dendrites, numerous different circuits can form endowing the nervous system with great plasticity, due to which it can respond to the numerous situations that may confront the organism.

Neuronal cell: See *Neuron.*

Nominalism: In the debate on the species, which has never abated, nominalism asserts that only individuals really exist. The nominalist does not deny that certain individuals that are classified in the same species resemble each other but this resemblance does not arise from a constitutive principle, or essence, inherent in the individuals. In contrast, for realists the species is perfectly real. It is a structure shared by several individuals, and is the result of a constitutive principle common to those individuals (see *Essentialism*).

Nucleosome: The nucleosome is the basic unit in chromatin fibre. It is a 'bead' of proteins, called histones, around which the DNA is wrapped. Chromatin fibre is composed of a string of nucleosomes, like a pearl necklace where each pearl is a nucleosome (see *Chromatin*).

Ontology: Ontology is the area concerned with indemonstrable first principles. In discussions in the realm of ontology, one is often induced to say that an entity is not real, for example, species or the individual organism. This means that it is not a first principle constituting what is real. For example, for a nominalist, the species is not real in that it depends on subjective classification by a classifier.

Phenotype: All the qualitative (e.g. eye colour) or quantitative (e.g. size) characteristics of an individual resulting from the expression of its genes.

Phosphatase: Post-translational modification enzyme dephosphorylating (removing phosphate from) proteins (see *Protein synthesis*).

Post-genomic biology: This term refers to the research programme that has been developing since sequencing the human genome. It is not strictly defined. One of its aspects consists in vastly widening programmes for collating data on proteins and RNAs (to include proteomes and transcriptomes) and in devising computer tools to analyse them. This modelling more and more often involves the participation of mathematicians and physicists.

Post-translational modifications: See *Protein synthesis*.

Prokaryote: Simple cell without a nucleus delimited by a membrane, e.g. bacteria.

Promoter: DNA sequence situated upstream of the genes where the enzyme allowing their transcription binds (See *Protein synthesis*).

Protein synthesis: The nucleotide sequence (the genetic information) of DNA is first of all transcribed into RNA (transcription). This

RNA is then itself translated into a linear chain of amino acids (translation). This chain in turn folds to form a three-dimensional structure. Finally, in general under the action of enzymes, this three-dimensional structure undergoes what are called post-translational or epigenetic modifications corresponding to chemical modifications of the amino acids of the protein. According to current theory, the properties of a protein depend on its three-dimensional structure which, through its form and electrical charges, permits interactions with other molecules (see *Molecular interaction*).

Proteome: All the proteins (and their interactions) of a cell.

Real: See *Ontology.*

Regulator protein: Protein in the chromatin which activates or inhibits gene transcription (see *Chromatin*, *Molecular interaction*).

Regulatory region: See *Regulatory sequence*, *Promoter.*

Regulatory sequence: DNA sequence situated upstream of the genes where the regulatory factors which activate or inhibit gene transcription bind (see *Protein synthesis*).

Specificity: That which exclusively identifies a series of things or organisms which thus constitute a species. However, the word 'specificity' is one of the most used terms in biological literature, employed with a variety of meanings which are not always explained. When we speak of molecular specificity, we are using the term in the original meaning of 'stereospecificity' defined by molecular biologists. Stereospecificity (etymology: solid specificity) indicates that molecules are only capable of performing unequivocal interactions, or very few in number, determined by their three-dimensional structure. Each molecule has only a single partner molecule (or a very restricted number of them) with which it can interact, which excludes random interactions between molecules.

Stem cells: Non-differentiated cells in the embryo or the adult organism that can either renew themselves by multiplication or differentiate.

Stereospecificity: See *Specificity.*

Synapse: See *Neuron.*

Teleonomy: Property of organisms of being endowed with an objective that they realise due to the genetic programme. This is a modern version of finality in the context of molecular biology.

Thermal agitation: The atoms and molecules in matter, irrespective of its state, are perpetually in motion, never immobile. This continual movement is correlated with the temperature: it reduces as the temperature falls (and totally ceases at absolute zero) and increases as the temperature rises. Although subject to Newton's deterministic laws, this movement of atoms or molecules cannot be predicted other than in a probabilistic way, as for the movement of a coin in the game of heads or tails. Molecules diffuse by this random Brownian motion which in the absence of constraints tends to make their concentrations homogeneous (see *Molecular interaction*).

Three-dimensional structure of proteins: See *Protein synthesis.*

Transcription: See *Protein synthesis.*

Translation: See *Protein synthesis.*

List of Abbreviations

CN	Chance and necessity (Jacques Monod)
CF	*Entre le cristal et la fumée* (Henri Atlan)
EE	Emergent evolution (Conrad LLoyd Morgan)
EH	Essays upon heredity and kindred biological problems (August Weismann)
FTG	*La fin du "tout génétique"*? (Henri Atlan)
GA	On the generation of animals (Aristotle)
HE	Holism and evolution (Jan Christiaan Smuts)
HU	At home in the universe (Stuart Kauffman)
ISEM	An introduction to the study of experimental medicine (Claude Bernard)
LPL	Lectures on the phenomena of life common to animals and plants (Claude Bernard)
OG	On generation (Hippocrates)
ONC	On the nature of the child (Hippocrates)
OOC	Order out of chaos (Ilya Prigogine and Isabelle Stengers)
OS	The origin of species (Charles Darwin)
PA	Parts of animals (Aristotle)
RTO	Reflections on a theory of organisms: Holism in biology (Walter Elsasser)
SL	The science of life (Paul Weiss)
WIL	What is life? (Erwin Schrödinger)

List of Figures

Bibliography

Albert R. (2005). Scale-free networks in cell biology. *J Cell Sci* **118**: 4947–4957.

Alexander S. (1920). *Space, Time and Deity.* Macmillan, London.

Ames P, Studdert CA, Reiser RH, Parkinson JS. (2002). Collaborative signaling by mixed chemoreceptor teams in *Escherichia coli. Proc Natl Acad Sci USA* **99**: 7060–7065.

Amrani A, Serra P, Yamanouchi J, Trudeau JD, Tan R, Elliott JF, Santamaria P. (2001). Expansion of the antigenic repertoire of a single T cell receptor upon T cell activation. *J Immunol* **167**: 655–666.

Arias AM, Hayward P. (2006). Filtering transcriptional noise during development: concepts and mechanisms. *Nat Rev Genet* **7**: 34–44.

Aristotle. *Physics.* Translated by Robin Waterfield. Oxford University Press, USA.

Aristotle. *On the Generation of Animals.* Translated by Arthur Platt. Large type edition, www.ReadHowYouWant.com, USA.

Aristotle. *Parts of Animals.* Translated by A. L. Peck. Harvard University Press, USA.

Atamas SP. (1996). Self-organization in computer simulated selective systems. *Biosystems* **39**: 143–151.

Atamas SP. (2003). Hasard et dégénérescence [Chance and Degeneracy]. *Sciences et Avenir Hors-série* **136**: 34–38.

Arthur W. (1988). *A Theory of the Evolution of Development.* Wiley, New York.

Atlan H. (1972, republished 2006). *L'organisation biologique et la théorie de l'information.* Seuil, Paris.

Atlan H. (1979, republished 1986). *Entre le cristal et la fumée* [*Between Crystal and Smoke*]. Points/Sciences, Seuil, Paris.

Atlan H. (1999). *La fin du "tout génétique"? [The End of all Genetics?]* INRA Editions.

Barabási AL, Oltvai ZN. (2004). Network biology: understanding the cell's functional organization. *Nat Rev Genet* **5**: 101–113.

Baroffio A, Blot M. (1992). Statistical evidence for a random commitment of pluripotent cephalic neural crest cells. *J Cell Sci* **103**(Pt 2): 581–587.

Beckett D. (2004). Functional switches in transcription regulation; molecular mimicry and plasticity in protein-protein interactions. *Biochemistry* **43**: 7983–7991.

Becskei A, Séraphin B, Serrano L. (2001). Positive feedback in eukaryotic gene networks: cell differentiation by graded to binary response conversion. *EMBO Journal* **20**: 2528–2535.

Bencharit S, Morton CL, Potter PM, Rebindo MR. (2003). Structural basis of heroin and cocaine metabolism by a promiscuous human drug-processing enzyme. *Nat Struct Biol* **10**: 349–356.

Bendall AJ, Sturm RA, Danoy PA, Molloy PL. (1993). Broad binding-site specificity and affinity properties of octamer 1 and brain octamer-binding proteins. *Eur J Biochem* **217**: 799–811.

Ben-Jacob E, Shmueli H, Schochet O, Tenenbaum A. (1992). Adapatative self-organization during growth of bacterial colonies. *Physica A* **187**: 378–424.

Bennett DC. (1983). Differentiation in mouse melanoma cells: initial reversibility and an on-off stochastic model. *Cell* **34**: 445–453.

Berg OG, von Hippel PH. (1985). Diffusion-controlled macromolecular interactions. *Annu Rev Biophys Biophys Chem* **4**: 131–160.

Bergson H. (1907, English translation 2007). *Creative Evolution.* Palgrave Macmillan.

Bernard C. (1865, English translation 1957). *An Introduction to the Study of Experimental Medicine.* Translated by Henry C. Greene. Dover, USA.

Bernard C. (1878, English translation 1974). *Lectures on the Phenomena of Life common to animals and plants.* Translated by H. E. Hoff, R. Guillemin and L. Guillemin. Charles C. Thomas, USA.

Biggin MD. (2001). To bind or not to bind. *Nat Genet* **28**: 303–304.

Bissell MJ, Weaver VM, Lelièvre SA, Wang F, Petersen OW, Schmeichel KL. (1999). Tissue structure, nuclear organization, and gene expression in normal and malignant breast. *Cancer Res* **59**: 1757–1763.

Blake WJ, Kaern M, Cantor CR, Collins JJ. (2003). Noise in eukaryotic gene expression. *Nature* **422**: 633–637.

Böhme K, Winterhalter KH, Bruckner P. (1995). Terminal differentiation of chondrocytes in culture is a spontaneous process and is arrested by transforming growth factor-beta 2 and basic fibroblast growth factor in synergy. *Exp Cell Res* **216**: 191–198.

Bork P, Jensen LJ, von Mering C, Ramani AK, Lee I, Marcotte EM. (2004). Protein interaction networks from yeast to human. *Curr Opin Struct Biol* **14**: 292–299.

Bouck GB, Brown DL. (1976). Self-assembly in development. *Annu Rev Plant Physiol* **27**: 71–94.

Bourbon HM, Martin-Blanco E, Rosen D, Kornberg TB. (1995). Phosphorylation of the Drosophila engrailed protein at a site outside its homeodomain enhances DNA binding. *J Biol Chem* **270**: 11130–11139.

Bouwmeester T. (2001). The Spemann-Mangold organizer: the control of fate specification and morphogenetic rearrangements during gastrulation in Xenopus. *Int J Dev Biol* **45**: 251–258.

Brandon RN, Burian RM (eds.). (1984). *Genes, Organisms, Populations: Controversies Over the Units of Selection.* MIT Press, Cambridge, MA.

Bray D. (2003). Molecular prodigality. *Science* **299**: 1189–1190.

Britten RJ. (1998). Underlying assumptions of developmental models. *Proc Natl Acad Sci USA* **95**: 9372–9377.

Broxmeyer HE, Kim CH. (1999). Regulation of hematopoiesis in a sea of chemokine family members with a plethora of redundant activities. *Exp Hematol* **27**: 1113–1123.

Burian R. (2008). On microRNA and the need for exploratory experimentation in post-genomic molecular biology. *Hist Philos Life Sci* **29**: 289–310.

Buss LW. (1987). *The Evolution of Individuality.* Princeton University Press, Princeton, NJ.

Camazine S, Deneubourg JL, Franks NR, Sneyd J, Theraulaz G, Bonabeau E. (2003). *Self-Organization in Biological Systems.* Princeton University Press, Princeton, USA.

Campbell SL, Khosravi-Far R, Rossman KL, Clark GJ, Der CJ. (1998). Increasing complexity of Ras signaling. *Oncogene* **17**: 1395–1413.

Capp JP. (2005). Stochastic gene expression, disruption of tissue averaging effects and cancer as a disease of development. *Bioessays* **27**: 1277–1285.

Capp JP. (In press). Le rôle de l'expression stochastique des gènes dans la genèse du cancer. In Gandrillon O, Kupiec JJ, Morange M (eds.), *Le hasard au cœur de la cellule.* Syllepse, Paris.

Carr A, Biggin MD. (1999). A comparison of *in vivo* and *in vitro* DNA-binding specificities suggests a new model for homeoprotein DNA binding in Drosophila embryos. *EMBO Journal* **18**: 1598–1608.

Carpenter G. (2000). The EGF receptor: a nexus for trafficking and signalling. *Bioessays* **22**: 697–707.

Caspar DLD, Klug A. (1962). Physical principles in the construction of regular viruses. *Cold Spring Harb Symp Quant Biol* **27**: 1–24.

Castagnoli L, Costantini A, Dall'Armi C, Gonfloni S, Montecchi-Palazzi L, Panni S, Paoluzi S, Santonico E, Cesareni G. (2004). Selectivity and promiscuity in the interaction network mediated by protein recognition modules. *FEBS Letters* **567**: 74–79.

Changeux JP, Courrège P, Danchin A. (1973). A theory of the epigenesis of neuronal networks by selective stabilization of synapses. *Proc Natl Acad Sci USA* **70**: 2974–2978.

Chess A, Simon I, Cedar H, Axel R. (1994). Allelic inactivation regulates olfactory receptor gene expression. *Cell* **78**: 823–834.

Child CM. (1941). *Patterns and Problems of Development.* The University of Chicago Press, Chicago.

Chubb JR, Bickmore WA. (2003). Considering nuclear compartmentalization in the light of nuclear dynamics. *Cell* **112**: 403–406.

Coulon A, Beslon G, Gandrillon O. (In press). Le hasard au cœur du noyau: comment la machinerie moléculaire régule la stochasticité de l'expression des gènes. In Gandrillon O, Kupiec JJ, Morange M (eds.), *Le hasard au cœur de la cellule.* Syllepse, Paris.

Cremer T, Cremer C. (2001). Chromosome territories, nuclear architecture and gene regulation in mammalian cells. *Nat Rev Genet* **2**: 292–301.

Crick FHC. (1958). On protein synthesis. *Symp Soc Exp Biol* **XII**: 139–163.

D'Ari R, Casadesus J. (1998). Underground metabolism. *Bioessays* **20**: 181–186.

Darwin C. (1859, republished 1985). *On the Origin of Species* (1st edition). Penguin (Penguin Classics).

Darwin, CR. (1868, republished 1998). *The Variation of Animals and Plants under Domestication.* John Hopkins University Press.

Davis CB, Killeen N, Crooks ME, Raulet D, Littman DR. (1993). Evidence for a stochastic mechanism in the differentiation of mature subsets of T lymphocytes. *Cell* **73**: 237–247.

Dawkins R. (1976, republished 2006). *The Selfish Gene.* Oxford University Press, Oxford.

de Laat W, Grosveld F. (2007). Inter-chromosomal gene regulation in the mammalian cell nucleus. *Curr Opin Genet Dev* **17**: 456–464.

Delbrück M. (1971). Aristotle-totle-totle. In Monod, Barek J (eds.), *Of Microbes and Life*, pp. 50–55. Columbia University Press, New York.

Díaz B, Moreno E. (2005). The competitive nature of cells. *Exp Cell Res* **306**: 317–322.

Dillon N, Festenstein R. (2002). Unravelling heterochromatin: competition between positive and negative factors regulates accessibility. *Trends Genet* **18**: 252–258.

Dinsmore (ed.). *A History of Regeneration Research. Milestone in the Evolution of Science*, pp. 201–217. Cambridge University Press, Cambridge.

Dobzhansky T. (1970). *Genetics of the Evolutionary Process.* Columbia University Press, New York.

Driesch H. (1908). *The Science and Philosophy of the Organism.* Adam and Charles Black, London.

Driesch H. (1914). *The Problem of Individuality.* MacMillan, London.

Dumont JE, Dremier S, Pirson I, Maenhaut C. (2002). Cross signaling, cell specificity, and physiology. *Am J Physiol* (*Cell Physiol*) **283**: C2–28.

Dundr M, Misteli T. (2001). Functional architecture in the cell nucleus. *Biochem J* **356**(Pt 2): 297–310.

Dunker AK, Cortese MS, Romero P, Iakoucheva LM, Uversky VN. (2005). Flexible nets. The roles of intrinsic disorder in protein interaction networks. *FEBS Journal* **272**: 5129–5148.

Dunker AK, Obradovic Z. (2001). The protein trinity — linking function and disorder. *Nat Biotechnol* **19**: 805–806.

Dunker AK, Obradovic Z, Romero P, Garner EC, Brown CJ. (2000). Intrinsic protein disorder in complete genomes. *Genome Inform* **11**: 161–171.

Dutoit V, Rubio-Godoy V, Pittet MJ, Zippelius A, Dietrich PY, Legal FA, Guillaume P, Romero P, Cerottini JC, Houghten RA, Pinilla C, Valmori D. (2002). Degeneracy of antigen recognition as the molecular basis for the high frequency of naive A2/Melan-a peptide multimer(+) $CD8^+$ T cells in humans. *J Exp Med* **196**: 207–216.

Dyson HJ, Wright PE. (2005). Intrinsically unstructured proteins and their functions. *Nat Rev Mol Cell Biol* **6**: 197–208.

Edelman GM, Mountcastle VB. (1978). *The Mindful Brain.* MIT Press, Cambridge, MA.

Elowitz MB, Levine AJ, Siggia ED, Swain PS. (2002). Stochastic gene expression in a single cell. *Science* **297**: 1183–1186.

Elsasser W. (1998). *Reflections on a Theory of Organisms: Holism in Biology.* The John Hopkins University Press, Baltimore, USA.

Fiering S, Northrop JP, Nolan GP, Mattila PS, Crabtree GR, Herzenberg LA. (1990). Single cell assay of a transcription factor reveals a threshold in transcription activated by signals emanating from the T-cell antigen receptor. *Genes Dev* **4**: 1823–1834.

Fischer E. (1894). Einfluss der Configuration auf die Wirkung der Enzyme. *Berichte der deutschen chemischen Gesellschaft* **27**: 2985–2993.

Fortunel NO, Hatzfeld A, Hatzfeld JA. (2000). Transforming growth factor-beta: pleiotropic role in the regulation of hematopoiesis. *Blood* **96**: 2022–2036.

Fraser P, Bickmore W. (2007). Nuclear organization of the genome and the potential for gene regulation. *Nature* **447**: 413–417.

Fu W, Chen W. (2004). Roles of chemokines in thymopoiesis: redundancy and regulation. *Cell Mol Immunol* **1**: 266–273.

Garcia KC, Degano M, Pease LR, Huang M, Peterson PA, Teyton L, Wilson IA. (1998). Structural basis of plasticity in T cell receptor recognition of a self peptide-MHC antigen. *Science* **279**: 1166–1172.

Gayon J. (1992A). L'espèce sans la forme. In Gayon J, Wunenburger JJ (eds.), *Les figures de la forme: philosophie biologique et théorie esthétique*, pp. 49–61. L'Harmattan, Paris.

Gayon J. (1992B, English translation 1998). *Darwinism's Struggle for Survival: Heredity and the Hypothesis of Natural Selection.* Cambridge University Press.

Gayon J. (In press). Déterminisme génétique, déterminisme bernardien, déterminisme laplacien. In Gandrillon O, Kupiec JJ, Morange M (eds.), *Le hasard au cœur de la cellule.* Syllepse, Paris.

Gehring WJ, Qian YQ, Billeter M, Furukubo-Tokunaga K, Schier AF, Resendez-Perez D, Affolter M, Otting G, Wüthrich K. (1994). Homeodomain-DNA recognition. *Cell* **78**: 211–223.

Glisse J, Le Guillou L, Laforge B, Kupiec JJ. (In press). Le rôle de l'aléatoire dans le modèle darwinien de différenciation cellulaire. In Gandrillon O, Kupiec JJ, Morange M (eds.), *Le hasard au cœur de la cellule.* Syllepse, Paris.

Godsave SF, Slack JM. (1991). Single cell analysis of mesoderm formation in the Xenopus embryo. *Development* **111**: 523–530.

Gordon N. (1995). Apoptosis (programmed cell death) and other reasons for elimination of neurons and axons. *Brain Dev* **17**: 73–77.

Guggenmos J, Schubart AS, Ogg S, Andersson M, Olsson T, Mather IH, Linington C. (2004). Antibody cross-reactivity between myelin oligodendrocyte glycoprotein and the milk protein butyrophilin in multiple sclerosis. *J Immunol* **172**: 661–668.

Gurdon JB, Bourillot PY. (2001). Morphogen gradient interpretation. *Nature* **413**: 797–803.

Hacking I. (1975). *The Emergence of Probability.* Cambridge University Press, Cambrige, UK.

Haarmann CS, Green D, Casarotto MG, Laver DR, Dulhunty AF. (2003). The random-coil 'C' fragment of the dihydropyridine receptor II-III loop can activate or inhibit native skeletal ryanodine receptors. *Biochem J* **372**: 305–316.

Halford SE, Marko JF. (2004). How do site-specific DNA-binding proteins find their targets? *Nucleic Acids Res* **32**: 3040–3052.

Hanscombe O, Whyatt D, Fraser P, Yannoutsos N, Greaves D, Dillon N, Grosveld F. (1991). Importance of globin gene order for correct developmental expression. *Genes Dev* **5**: 1387–1394.

Hausmann S, Martin M, Gauthier L, Wucherpfennig KW. (1999). Structural features of autoreactive TCR that determine the degree of degeneracy in peptide recognition. *J Immunol* **162**: 338–344.

Haynes C, Oldfield CJ, Ji F, Klitgord N, Cusick ME, Radivojac P, Uversky VN, Vidal M, Iakoucheva LM. (2006). Intrinsic disorder is a common feature of hub proteins from four eukaryotic interactomes. *PLoS Comput Biol* **2**: e100.

Heams T, Kupiec JJ. (2003). Modified 3′-end amplification PCR for gene expression analysis in single cells. *Biotechniques* **34**: 712–716.

Heams T. (2004). *Approche endodarwinienne de la variabilité intercellulaire de l'expression génétique.* PhD thesis for INA-PG (director: Kupiec JJ).

Heams T. (In press). Expression stochastique des gènes et différenciation cellulaire. In Gandrillon O, Kupiec JJ, Morange M (eds.), *Le hasard au cœur de la cellule.* Syllepse, Paris.

Held W, Kunz B, Ioannidis V, Lowin-Kropf B. (1999). Mono-allelic Ly49 NK cell receptor expression. *Seminars in Immunology* **11**: 349–355.

Hill CS, Treisman R. (1995). Transcriptional regulation by extracellular signals: mechanisms and specificity. *Cell* **80**: 199–211.

Hippocrates. *The Hippocratic Treatises — On Generation, On the Nature of the Child,* Translated by I. M Lonie de Gruyter.

Holländer GA. (1999). On the stochastic regulation of interleukin-2 transcription. *Seminars in Immunology* **11**: 357–367.

Hull DL. (1982). Philosophy and Biology. In Fløistad G (ed.), *Contemporary Philosophy, A New Survey, Philosophy of Science*, vol. 2: pp. 280–316. Nijhoff.

Hume D. (1739, republished 2007). *A Treatise of Human Nature.* NuVision Publications.

Hunter T. (2000). Signaling — 2000 and beyond. *Cell* **100**: 113–127.

Huxley JS. (1942). *Evolution: The Modern Synthesis.* Allen and Unwin, Australia.

Iakoucheva LM, Brown CJ, Lawson JD, Obradović Z, Dunker AK. (2002). Intrinsic disorder in cell-signaling and cancer-associated proteins. *J Mol Biol* **323**: 573–584.

Inoué S. (1982). The role of self-assembly in the generation of biological form. In Subtelny S, Green PB (eds.), *Developmental order: its origin and regulation (40th symposium of the Society for Developmental Biology)* pp. 35–37.

Isaacs FJ, Hasty J, Cantor CR, Collins JJ. (2003). Prediction and measurement of an autoregulatory genetic module. *Proc Natl Acad Sci USA* **100**: 7714–7719.

Jablonka E, Lamb MJ. (1995). *Epigenetic Inheritance and Evolution: The Lamarckian Dimension.* Oxford University Press, Oxford.

Jacob F, Monod J. (1961). Genetic regulatory mechanisms in the synthesis of proteins. *J Mol Biol* **3**: 318–356.

Jacob F. (1970, English translation 1993). *The Logic of Life.* Princeton University Press.

Jerne NK. (1955). The natural-selection theory of antibody formation. *Proc Natl Acad Sci USA* **41**: 849–857.

Jouvenot Y, Poirier F, Jami J, Paldi A. (1999). Biallelic transcription of Igf2 and H19 in individual cells suggests a post-transcriptional contribution to genomic imprinting. *Curr Biol* **9**: 1199–1202.

Jungermann K, Sasse D. (1978). Heterogeneity of liver parenchymal cells. *Trends Biochem Sci* **3**: 198–202.

Kaern M, Elston TC, Blake WJ, Collins JJ. (2005). Stochasticity in gene expression: from theories to phenotypes. *Nat Rev Genet* **6**: 451–464.

Kaufmann BB, van Oudenaarden A. (2007). Stochastic gene expression: from single molecules to the proteome. *Curr Opin Genet Dev* **17**: 107–112.

Kauffman SA. (1993). *The Origins of Order: Self-Organization and Selection in Evolution.* Oxford University Press, Oxford.

Kauffman SA. (1995). *At Home in the Universe.* Oxford University Press, Oxford.

Kazlauskas A. (1994). Receptor tyrosine kinases and their targets. *Curr Opin Genet Dev* **4**: 5–14.

Kendrew JC, Bodo G, Dintzis H, Parrish RG, Wyckoff HW, Phillips DC. (1958). A three-dimensional model of the myoglobin molecule obtained by X-ray analysis. *Nature* **181**: 662–666.

Kenny PA, Bissell MJ. (2003). Tumor reversion: correction of malignant behavior by microenvironmental cues. *Int J Cancer* **107**: 688–695.

Kepler TB, Elston TC. (2001). Stochasticity in transcriptional regulation: origins, consequences, and mathematical representations. *Biophys J* **81**: 3116–3136.

Kim J. (2002). The layered model: metaphysical considerations. *Philosophical Explorations* **5**: 2–20.

Kirschner M, Gerhart J, Mitchison T. (2000). Molecular "vitalism". *Cell* **100**: 79–88.

Kitano H. (2002). Systems biology: a brief overview. *Science* **295**: 1162–1164.

Ko MS, Nakauchi H, Takahashi N. (1990). The dose dependence of glucocorticoid-inducible gene expression results from changes in the number of transcriptionally active templates. *EMBO Journal* **9**: 2835–2842.

Komarova NL, Zou X, Nie Q, Bardwell L. (2005). A theoretical framework for specificity in cell signaling. *Mol Syst Biol* **1**: 2005.0023.

Krause R, von Mering C, Bork P, Dandekar T. (2004). Shared components of protein complexes — versatile building blocks or biochemical artefacts? *Bioessays* **26**: 1333–1343.

Kupiec JJ. (1981). Théorie probabiliste de la différenciation cellulaire. *XIIème Rencontres de Méribel*, pp. 161–163.

Kupiec JJ. (1983). A probabilist theory for cell differentiation, embryonic mortality and DNA C-value paradox. *Speculation in Science and Technology* **6**: 471–478.

Kupiec JJ. (1986). A probabilist theory for cell differentiation: the extension of Darwinian principles to embryogenesis. *Speculation in Science and Technology* **9**: 19–22.

Kupiec JJ. (1989). Gene regulation and DNA C-value paradox: a model based on diffusion of regulatory molecules. *Med Hypotheses* **28**: 7–10.

Kupiec JJ. (1996). A chance-selection model for cell differentiation. *Cell Death Differ* **3**: 385–390.

Kupiec JJ. (1997). A Darwinian theory for the origin of cellular differentiation. *Mol Gen Genet* **255**: 201–208.

Kupiec JJ. (1999). L'influence de la philosophie d'Aristote sur l'élaboration de la théorie de l'évolution et sur la génétique. *Revue Européenne de Sciences Sociales* **XXXVII(**115): 89–116.

Kupiec JJ, Sonigo P. (2000). *Ni Dieu ni gène.* Seuil, Paris.

Kupiec JJ. (2001). Retour vers le phénotype — les gènes existent-ils? In Brohm JM (ed.), *Le Vivant, Prétentaine,* pp. 263–266. University of Montpellier III.

Kupiec JJ. (2004). L'hérédité entre l'amortalité du corps et l'immortalité de l'âme. In Lenoir F, de Tonnac JP (eds.), *La mort et l'immortalité,* pp. 1485–1504. Bayard, Paris.

Kupiec JJ. (2005). Un ordre humain, trop humain. *Sciences et Avenir Hors-série* **143**: 30–34.

Laforge B. (2004). Prolégomènes à une nouvelle théorie du vivant. In Dubessy J, Lecointre G, Silberstein M (eds.), *Les matérialismes (et leurs détracteurs)*, pp. 263–278. Syllepse, Paris.

Laforge B, Guez D, Martinez M, Kupiec JJ. (2005). Modeling embryogenesis and cancer: an approach based on an equilibrium between the autostabilization of stochastic gene expression and the interdependence of cells for proliferation. *Prog Biophys Mol Biol* **89**: 93–120.

Lefebvre P, Gaub MP, Tahayato A, Rochette-Egly C, Formstecher P. (1995). Protein phosphatases 1 and 2A regulate the transcriptional and DNA binding activities of retinoic acid receptors. *J Biol Chem* **270**: 10806–10816.

Lenoble R. (1969). *Histoire de l'idée de nature.* Albin Michel, Paris.

Levchenko A. (2003). Dynamical and integrative cell signaling: challenges for the new biology. *Biotechnol Bioeng* **84**: 773–782.

Lewis J. (1998). Notch signalling and the control of cell fate choices in vertebrates. *Semin Cell Dev Biol* **9**: 583–589.

Lewis J, Slack JM, Wolpert L. (1977). Thresholds in development. *J Theor Biol* **65**: 579–590.

Li CC, Dai RM, Chen E, Longo DL. (1994). Phosphorylation of NF-KB1-p50 is involved in NF-kappa B activation and stable DNA binding. *J Biol Chem* **269**: 30089–30092.

Lin Z, Lu MH, Schultheiss T, Choi J, Holtzer S, DiLullo C, Fischman DA, Holtzer H. (1994). Sequential appearance of muscle-specific proteins in myoblasts as a function of time after cell division: evidence for a conserved myoblast differentiation program in skeletal muscle. *Cell Motil Cytoskeleton* **29**: 1–19.

Liu J, Perumal NB, Oldfield CJ, Su EW, Uversky VN, Dunker AK. (2006). Intrinsic disorder in transcription factors. *Biochemistry* **45**: 6873–6888.

Longo G, Tendero PE. (2007). The differential method and the causal incompleteness of Programming Theory in Molecular Biology. *Foundations of Science* **12**: 337–366.

Logan CY, Nusse R. (2004). The Wnt signaling pathway in development and disease. *Annu Rev Cell Dev Biol* **20**: 781–810.

Ma B, Shatsky M, Wolfson HJ, Nussinov R. (2002). Multiple diverse ligands binding at a single protein site: a matter of pre-existing populations. *Protein Science* **11**: 184–197.

Maamar H, Raj A, Dubnau D. (2007). Noise in gene expression determines cell fate in Bacillus subtilis. *Science* **317**: 526–529.

Maffini MV, Soto AM, Calabro JM, Ucci AA, Sonnenschein C. (2004). The stroma as a crucial target in rat mammary gland carcinogenesis. *J Cell Sci* **117**(Pt 8): 1495–1502.

Manivel V, Sahoo NC, Salunke DM, Rao KV. (2000). Maturation of an antibody response is governed by modulations in flexibility of the antigen-combining site. *Immunity* **13**: 611–620.

Manivel V, Bayiroglu F, Siddiqui Z, Salunke DM, Rao KVS. (2002). The primary antibody repertoire represents a linked network of degenerate antigen specificities. *J Immunol* **169**: 888–897.

Martin T. (In press). De la diversité des probabilités. In Gandrillon O, Kupiec JJ, Morange M (eds.), *Le hasard au cœur de la cellule.* Syllepse, Paris.

Mauron A. (2002). Genomic metaphysics. *J Mol Biol* **319**: 957–962.

Maynard Smith J, Burian R, Kaufman S, Alberch P, Campbell J, Goodwin B, Lande R, Raup D, Wolpert L. (1985). Development constraints and evolution. *Q Rev Biol* **60**: 265–287.

Mayr E. (1961). Cause and effect in biology. *Science* **134**: 1501–1506.

Mayr E. (1982). *The Growth of Biological Thought: Diversity, Evolution, and Inheritance.* Harvard University Press, Cambridge, MA.

Mayr E. (1993). *One Long Argument.* Harvard University Press, Cambridge, MA.

Mayr E, Provine WB (eds.). (1998). *The Evolutionary Synthesis: Perspectives on the Unification of Biology.* Harvard University Press, Cambridge, MA.

McAdams HH, Arkin A. (1997). Stochastic mechanisms in gene expression. *Proc Natl Acad Sci USA* **94**: 814–819.

McAdams HH, Arkin A. (1999). It's a noisy business! Genetic regulation at the nanomolar scale. *Trends Genet* **15**: 65–69.

McGrane MM, Yun JS, Patel YM, Hanson RW. (1992). Metabolic control of gene expression: *in vivo* studies with transgenic mice. *Trends Biochem Sci* **17**: 40–44.

Merlin F. (In press). Pour une interprétation objective des probabilités dans les modèles d'expression stochastique des gènes. In Gandrillon O, Kupiec JJ, Morange M (eds.), *Le hasard au cœur de la cellule.* Syllepse, Paris.

Meshorer E, Yellajoshula D, George E, Scambler PJ, Brown DT, Misteli T. (2006). Hyperdynamic plasticity of chromatin proteins in pluripotent embryonic stem cells. *Dev Cell* **10**: 105–116.

Michaelson J. (1993). Cellular selection in the genesis of multicellular organization. *Lab Invest* **69**: 136–151.

Mirsky AE, Pauling L. (1936). On the structure of native, denatured and coagulated proteins. *Proc Natl Acad Sci USA* **22**: 439–447.

Misteli T. (2001). Protein dynamics: implications for nuclear architecture and gene expression. *Science* **291**: 843–847.

Misteli T. (2007). Beyond the sequence: cellular organization of genome function. *Cell* **128**: 787–800.

Moggs JG, Orphanides G. (2001). Estrogen receptors: orchestrators of pleiotropic cellular response. *EMBO Reports* **21**: 775–781.

Monod J, Jacob F. (1961). General conclusions: Teleonomic mechanisms in cellular metabolism, growth and differentiation. *Cold Spring Harb Symp Quant Biol* **XXVI**: 487–499.

Monod J. (1970, English translation 1977). *Chance and Necessity.* Collins/ Fount paperbacks.

Morange M. (1994, English translation 2000). *A History of Molecular Biology.* Harvard University Press.

Moreno E, Basler K, Morata G. (2002). Cells compete for decapentaplegic survival factor to prevent apoptosis in Drosophila wing development. *Nature* **416**: 755–759.

Moreno E, Basler K. (2004). dMyc transforms cells into super-competitors. *Cell* **117**: 117–129.

Morgan CL. (1923). *Emergent Evolution.* Henry Holt and Company, New York.

Morgan TH. (1926). *The Theory of the Gene.* Yale University Press, New Haven.

Morgan TH. (1934, republished 1998). *Embryology and Genetics.* Agrobios, India.

Mundorff EC, Hanson MA, Varvak A, Ulrich H, Schultz PG, Stevens RC. (2000). Conformational effects in biological catalysis: an antibody-catalysed Oxy-Cope rearrangement. *Biochemistry* **39**: 627–632.

Nakahata T, Gross AJ, Ogawa M. (1982). A stochastic model of self-renewal and commitment to differentiation of the primitive hemopoietic stem cells in culture. *J Cell Physiol* **113**: 455–458.

Nan X, Campoy FJ, Bird A. (1997). MeCP2 is a transcriptional repressor with abundant binding sites in genomic chromatin. *Cell* **88**: 471–481.

Neildez-Nguyen TM, Parisot A, Vignal C, Rameau P, Stockholm D, Picot J, Allo V, Le Bec C, Laplace C, Paldi A. (2008). Epigenetic gene expression noise and phenotypic diversification of clonal cell populations. *Differentiation* **76**: 33–40.

Nemazee D. (2000). Receptor selection in B and T lymphocytes. *Annu Rev Immunol* **18**: 19–51.

Newlands S, Levitt LK, Robinson CS, Karpf AB, Hodgson VR, Wade RP, Hardeman EC. (1998). Transcription occurs in pulses in muscle fibers. *Genes Dev* **12**: 2748–2758.

Noble D. (2006). *The Music of Life.* Oxford University Press, Oxford.

Noble D. (2008). Principes de la biologie des systèmes et leurs applications en modélisation. *Matière Première* (*Syllepse*) **3**: 239–252.

Ohgiwari M, Matsushita M, Matsuyama T. (1992). Morphological changes in growth phenomena of bacterial colony patterns. *J Physical Soc Japan* **61**: 816–822.

Paldi A. (2003). Stochastic gene expression during cell differentiation: order from disorder? *Cell Mol Life Sci* **60**: 1775–1778.

Penaloza C, Lin L, Lockshin RA, Zakeri Z. (2006). Cell death in development: shaping the embryo. *Histochem Cell Biol* **126**: 149–158.

Paulus U, Loeffler M, Zeidler J, Owen G, Potten CS. (1993). The differentiation and lineage development of goblet cells in the murine small intestinal crypt: experimental and modelling studies. *J Cell Sci* **106**(Pt 2): 473–483.

Pawson T, Nash P. (2003). Assembly of cell regulatory systems through protein interaction domains. *Science* **300**: 445–452.

Pichot A. (1993). *Histoire de la notion de vie.* Gallimard, Paris.

Pittendrigh CS. (1958). Adaptation, natural selection, and behavior. In Roe A, Simpson GG (eds.), *Behavior and Evolution*, pp. 390–416. Yale University Press, New Haven, CT.

Popper K. (1992). *Un univers de propensions.* Editions de l'Eclat, Paris.

Prigogine I, Stengers I. (1979, English edition 1985). *Order Out of Chaos.* Fontana paperbacks.

Prill RJ, Iglesias PA, Levchenko A. (2005). Dynamic properties of network motifs contribute to biological network organization. *PLoS Biology* **3**: e343.

Raser JM, O'Shea EK. (2004). Control of stochasticity in eukaryotic gene expression. *Science* **30**: 1811–1814.

Raser JM, O'Shea EK. (2005). Noise in gene expression: origins, consequences, and control. *Science* **309**: 2010–2013.

Rivière I, Sunshine MJ, Littman DR. (1998). Regulation of IL-4 expression by activation of individual alleles. *Immunity* **9**: 217–228.

Robb L. (2007). Cytokine receptors and hematopoietic differentiation. *Oncogene* **26**: 6715–6723.

Ross IL, Browne CM, Hume DA. (1994). Transcription of individual genes in eukaryotic cells occurs randomly and infrequently. *Immunol Cell Biol* **72**: 177–185.

Roux W. (1881, republished 2007). *Der Kampf der Teile im Organismus* [*The Struggle of Parts in the Organism*]. VDM, Saarbrücken, Germany.

Rubin H. (1990). On the nature of enduring modifications induced in cells and organisms. *Am J Physiol* **258**(2 Pt 1): L19–24.

Saha M. (1991). Spemann seen through a lens. In Gilbert SF (ed.), *A conceptual history of modern embryology*, pp. 91–108. Plenum Press, New York.

Schrödinger E. (1944, republished 1992). *What is Life?* (*with "Mind and Matter" and "Autobiographical Sketches"*). Cambridge University Press (Canto).

Schwartz MA, Madhani HD. (2004). Principles of MAP kinase signaling specificity in Saccharomyces cerevisiae. *Annu Rev Genet* **38**: 725–748.

Schweitzer R, Shilo BZ. (1997). A thousand and one roles for the Drosophila EGF receptor. *Trends Genet* **13**: 191–196.

Segal J. (2003). *Le zero et le un.* Syllepse, Paris.

Shapiro JA. (1995). The significances of bacterial colony patterns. *Bioessays* **17**: 597–607.

Smuts JC. (1926, republished 1999). *Holism and Evolution.* Sierra Sunrise Books, Sherman Oaks, Ca.

Socolovsky M, Fallon AE, Lodish HF. (1998). The prolactin receptor rescues EpoR-/- erythroid progenitors and replaces EpoR in a synergistic interaction with c-kit. *Blood* **92**: 1491–1496.

Sonigo P, Stengers I. (2003). *L'évolution.* EDP Sciences.

Sonnenschein CS, Soto AM. (1999). The society of cells. BIOS Scientific Publishers, Oxford.

Soto AM, Sonnenschein C. (2006). Emergentism by default: a view from the bench. *Synthese* **151**: 361–376.

Spemann H. (1938). *Embryonic Development and Induction.* Yale University Press, New Haven, CT.

Sperling R, Francus T, Siskind GW. (1983). Degeneracy of antibody specificity. *J Immunol* **131**: 882–885.

Stockholm D, Neildez TM, Vignal C, Parisot A, Benchaouir R, Picot J, Rameau, P, Landini G, Laplace-Builhe C, Pàldi A. (In press). L'origine de l'hétérogénéité phénotypique des cellules clonales. In Gandrillon O, Kupiec JJ, Morange M (eds.), *Le hasard au cœur de la cellule.* Syllepse, Paris.

Tabata T, Takei Y. (2004). Morphogens, their identification and regulation. *Development* **131**: 703–712.

Takasuka N, White MR, Wood CD, Robertson WR, Davis JR. (1998). Dynamic changes in prolactin promoter activation in individual living lactotrophic cells. *Endocrinology* **139**: 1361–1368.

Takenaka I, Morin F, Seizinger BR, Kley N. (1995). Regulation of the sequence-specific DNA binding function of p53 by protein kinase C and protein phosphatases. *J Biol Chem* **270**: 5405–5411.

Till JE, McCulloch EA, Siminovitch L. (1964). A stochastic model of stem cell proliferation, based on the growth of spleen colony-forming cells. *Proc Natl Acad Sci USA* **51**: 29–36.

Till JE. (1981). Cellular diversity in the blood-forming system. *Am Sci* **69**: 522–527.

Tjwa M, Luttun A, Autiero M, Carmeliet P. (2003). VEGF and PlGF: two pleiotropic growth factors with distinct roles in development and homeostasis. *Cell Tissue Res* **314**: 5–14.

Townes TM, Behringer RR. (1990). Human globin locus activation region (LAR): role in temporal control. *Trends Genet* **6**: 219–223.

Turing AM. (1952). The chemical basis of morphogenesis. *Philos Trans R Soc Lond B* **327**: 37–72.

van Kempen LC, Ruiter DJ, van Muijen GN, Coussens LM. (2003). The tumor microenvironment: a critical determinant of neoplastic evolution. *Eur J Cell Biol* **82**: 539–548.

van Roon MA, Aten JA, van Oven CH, Charles R, Lamers WH. (1989). The initiation of hepatocyte-specific gene expression within embryonic hepatocytes is a stochastic event. *Dev Biol* **136**: 508–516.

Vinci T, Robert JS. (2005). Aristotle and modern genetics. *J Hist Ideas* **66**: 201–221.

Von Foerster H. (1960). On Self-Organizing Systems and their Environments. In Yovitts MC, Cameron S (eds.), *Self-Organizing Systems*, Pergamon Press, pp. 31–50.

von Mering C, Krause R, Snel B, Cornell M, Oliver SG, Fields S, Bork P. (2002). Comparative assessment of large-scale data sets of protein-protein interactions. *Nature* **417**: 399–403.

Vyas S, Béchade C, Riveau B, Downward J, Triller A. (2002). Involvement of survival motor neuron (SMN) protein in cell death. *Hum Mol Genet* **11**: 2751–2764.

Weisbuch G. (1999). L'approche de la biologie à l'aide de systèmes complexes adaptés. In Feltz B, Crommelinck et M, Goujon P (eds.), *Auto-organisation et émergence dans les sciences de la vie*, pp. 183–209. Ousia, Brussels.

Weismann A. (1891). *Essay on Heredity and Kindred Biological Problems.* Clarendon Press, Oxford.

Weiss PA. (1973). *The Science of Life.* Futura Publishing Company, USA.

Wernet MF, Mazzoni EO, Celik A, Duncan DM, Duncan I, Desplan C. (2006). Stochastic spineless expression creates the retinal mosaic for colour vision. *Nature* **440**: 174–180.

White MR, Masuko M, Amet L, Elliott G, Braddock M, Kingsman AJ, Kingsman SM. (1995). Real-time analysis of the transcriptional regulation of HIV and hCMV promoters in single mammalian cells. *J Cell Sci* **108**(Pt 2): 441–455.

Wijgerde M, Grosveld F, Fraser P. (1995). Transcription complex stability and chromatin dynamics *in vivo*. *Nature* **377**: 209–213.

Williams GC. (1992). *Natural Selection: Domains, Levels, and Challenges.* Oxford University Press, Oxford.

Wolf U. (1997). Identical mutations and phenotypic variation. *Hum Genet* **100**: 305–321.

Wolpert L. (1969). Positional information and the spatial pattern of cellular differentiation. *J Theor Biol* **25**: 1–47.

Wolpert L. (1989). Positional information revisited. *Development* **107**(Supplement): 3–12.

Wolpert L (1991). Morgan's ambivalence: a history of gradients and regeneration. In CE.

Wright PE, Dyson HJ. (1999). Intrinsically unstructured proteins: re-assessing the protein structure-function paradigm. *J Mol Biol* **293**: 321–331.

Wu H. (1931). Studies on Denaturation of Proteins. XIII. A Theory of Denaturation. *Chin J Physiol* **5**: 321–344.

Xu M, Sheppard KA, Peng CY, Yee AS, Piwnica-Worms H. (1994). Cyclin A/CDK2 binds directly to E2F-1 and inhibits the DNA-binding activity of E2F-1/DP-1 by phosphorylation. *Mol Cell Biol* **14**: 8420–8431.

Yamada Y, Sakurada K, Takeda Y, Gojo S, Umezawa A. (2007). Single-cell-derived mesenchymal stem cells overexpressing Csx/Nkx2.5 and GATA4 undergo the stochastic cardiomyogenic fate and behave like transient amplifying cells. *Exp Cell Res* **313**: 698–706.

Zhang C, Xuan Z, Otto S, Hover JR, McCorkle SR, Mandel G, Zhang MQ. (2006). A clustering property of highly-degenerate transcription factor binding sites in the mammalian genome. *Nucleic Acids Res* **34**: 2238–2246.

Author Index

Subject Index